KU-004-611

Contents

Acknowledgements

Firstly, my thanks to Lisa Tenzin-Dolma: a great friend and true inspiration. Without Lisa's gentle nudge, this book would never had been written. Thanks also to Jude Brooks and Hubble & Hattie for allowing me to write about my memories before I forget them, and for agreeing to share them with you, the reader. Thank you to all the lifelong friends I have made, past and present, at the UK Wolf Conservation Trust: you know who you are, and I value your support and friendship. Your amazing photographic skills and generosity mean that this book contains some wonderful photos of the wolves.

Sending respect into the ether for Roger Palmer and John Denness for teaching me the difference between dogs and wolves. Thanks also to Pia Gismondi and Caroline Elliott, and Tony Haighway from Wolf Watch UK for continuing to allow me access to socialised wolves with whom to interact. It keeps me sane and inspired having contact with these amazing creatures who allow us into their world.

Wildwood Trust Education Officer Anne Riddell also deserves a mention as she never complains about me turning up to ogle the pack, and chat wolves for hours.

Mostly, though, I'd like to say a huge thanks to all of the animals who come into my life. Some of them just visit, some stay for a while, but all of them teach me something important ... and leave pawprints on my heart.

This book is dedicated to –
Kenai, Kodiak, Mika, Mai, Mosi, Torak, Lunca, Alba, Latea, Duma, and Dakota. I owe you all so much.

And to all captive animals, I wish your lives could be different, and I hope one day mankind will change how it treats all animalkind.

Contact details for Toni Shelbourne
www.tellingtontouch.co.uk
Facebook page The Truth about Wolves & Dogs
Twitter@tonishelbourne
YouTube videos – search for The Truth about Wolves and Dogs.
There are four videos –

- One – Why I wrote the book
- Two – The wolf pack
- Three – Canines/humans
- Four – The art of canine communication

AM … VES

…ourne

Hubble & Hattie

Memoirs of a wolf handler

The Hubble & Hattie imprint was launched in 2009 and is named in memory of two very special Westie sisters owned by Veloce's proprietors.

Since the first book, many more have been added to the list, all with the same underlying objective: to be of real benefit to the species they cover, at the same time promoting compassion, understanding and respect between all animals (including human ones!)

All Hubble & Hattie publications offer ethical, high quality content and presentation, plus great value for money.

More books from Hubble & Hattie –

Among the Wolves: Memoirs of a wolf handler (Shelbourne)
Animal Grief: How animals mourn (Alderton)
Because this is our home ... the story of a cat's progress (Bowes)
Camper vans, ex-pats & Spanish Hounds: from road trip torescue – the strays of Spain (Coates & Morris)
Cat Speak: recognising & understanding behaviour Rauth-Widmann)
Charlie – The dog who came in from the wild (Tenzin-Dolma)
Clever dog! Life lessons from the world's most successful animal (O'Meara)
Complete Dog Massage Manual, The – Gentle Dog Care (Robertson)
Dieting with my dog: one busy life, two full figures ... and unconditional love (Frezon)
Dinner with Rover: delicious, nutritious meals for you and your dog to share (Paton-Ayre)
Dog Cookies: healthy, allergen-free treat recipes for your dog (Schöps)
Dog-friendly Gardening: creating a safe haven for you and your dog (Bush)
Dog Games – stimulating play to entertain your dog and you (Blenski)
Dog Relax – relaxed dogs, relaxed owners (Pilguj)
Dog Speak: recognising & understanding behaviour (Blenski)
Dogs on Wheels: travelling with your canine companion (Mort)
Emergency First Aid for dogs: at home and away Revised Edition (Bucksch)
Exercising your puppy: a gentle & natural approach – Gentle Dog Care (Robertson & Pope)
Fun and Games for Cats (Seidl)
Gymnastricks: Targeted muscle training for dogs (Mayer)
Helping minds meet – skills for a better life with your dog (Zulch & Mills)
Know Your Dog – The guide to a beautiful relationship (Birmelin)
Life skills for puppies – laying the foundation of a loving, lasting relationship (Zuch & Mills)
Living with an Older Dog – Gentle Dog Care (Alderton & Hall)
Miaow! Cats really are nicer than people! (Moore)
My cat has arthritis – but lives life to the full! (Carrick)
My dog has arthritis – but lives life to the full! (Carrick)
My dog has cruciate ligament injury – but lives life to the full! (Haüsler & Friedrich)
My dog has epilepsy – but lives life to the full! (Carrick)
My dog has hip dysplasia – but lives life to the full! (Haüsler & Friedrich)
My dog is blind – but lives life to the full! (Horsky)
My dog is deaf – but lives life to the full! (Willms)
My Dog, my Friend: heart-warming tales of canine companionship from celebrities and other extraordinary people (Gordon)
No walks? No worries! Maintaining wellbeing in dogs on restricted exercise (Ryan & Zulch)
Partners – Everyday working dogs being heroes every day (Walton)
Smellorama – nose games for dogs (Theby)
Swim to recovery: canine hydrotherapy healing – Gentle Dog Care (Wong)
A tale of two horses – a passion for free-will teaching (Gregory)
The Truth about Wolves and Dogs: dispelling the myths of dog training (Shelbourne)
Waggy Tails & Wheelchairs (Epp)
Walking the dog: motorway walks for drivers & dogs (Rees)
Winston ... the dog who changed my life (Klute)
You and Your Border Terrier – The Essential Guide (Alderton)
You and Your Cockapoo – The Essential Guide (Alderton)
Your dog and you – understanding the canine psyche (Garratt)

Photo credits

All photos are from the author's own collection or those of friends and family. Copyright and ownership remain in the possession of the original photographer, and the images are used with their kind permission.

For post publication news, updates and amendments relating to this book please visit www.hubbleandhattie.com/extras/HH4760

www.hubbleandhattie.com

First published in May 2015 by Veloce Publishing Limited, Veloce House, Parkway Farm Business Park, Middle Farm Way, Poundbury, Dorchester, Dorset, DT1 3AR, England. Fax 01305 250479/email info@hubbleandhattie.com/web www.hubbleandhattie.com
ISBN: 978-1-845847-60-9 UPC: 6-36847-04760-3
Readers with ideas for books about animals, or animal-related topics, are invited to write to the editorial director of Veloce Publishing at the above address. British Library Cataloguing in Publication Data - A catalogue record for this book is available from the British Library. Typesetting, design and page make-up all by Veloce Publishing Ltd on Apple Mac. Printed in India by Replika Press

Foreword

Very few people have the privilege of working hand-in-paw with wolves on a daily basis. Toni Shelbourne's first book, *The Truth about Wolves and Dogs – dispelling the myths of dogs training*, has already become a classic text that sheds light on previous misunderstandings. Her second book for Hubble & Hattie, *Among the Wolves – memoirs of a wolf handler*, is deeply personal. Through the chapters we meet the wolves who were in Toni's care during her ten years at the UK Wolf Conservation Trust, and we learn how she developed a very special bond with each of them. In this compelling memoir, vividly recounted, Toni allows us to see these fascinating animals through her eyes: the North American wolves, Kenai and Kodiak; the North American types, Duma and Dakota; Alba, Lunca and Latea, the Europeans; and the Canadian wolves, Torak, Mosi, Mai, and Mika.

Wolves possess a mystique that appears to be unique among species. Hated, feared and misunderstood by some elements of society; revered almost to the point of worship by others, they evoke some primal aspect within many of us. They retain their wild natures even when socialised from an early age, and Toni makes it clear that any relationship must take place on their terms. Respect and trust, so important to all harmonious interactions, is even more vital for personal safety when the other party has jaws powerful enough to deliver 1500 pounds (680kg) of pressure.

The eleven wolves in Toni's care were socialised with humans from infancy. These animals were very different to the sad, fearful, timid creatures that we see pacing back and forth in zoos: captives whose environments lack the necessary space, social stimuli and enrichment, making their lives emotionally, mentally, and physically poverty-stricken. The wolves Toni writes about were carefully nurtured so that they could lead natural lives whilst living in close proximity to humans. She raised them, handled and played with them, taught them that the presence of people was desirable as well as acceptable, and she nursed them through sickness and injury with a tender expertise that is extraordinarily moving to read about. The wolves not only accepted her into their world, they welcomed and developed powerful relationships with her.

The personality of each wolf shines out from the book's pages: mischievous, fun-loving, authoritative, bullying, playful, affectionate, gentle, destructive, adventurous – Toni describes in delightful detail which of the wolves display each of these traits, so that we feel that we know, love and respect them, too. The chapters are rich in descriptions that evoke powerful feelings: the hair-raising image of Alba taking Toni's entire head in his jaws in a muzzle-hold, and the soft-focus, heart-warming image of three wolf cubs nestled asleep against her side, in-between clambering over her and romping around, reveal a great deal about wolf nature. This is a book which paints in intense detail what it's like to be among wild souls, and gain their trust and friendship.

Among the wolves

Dogs retain 99.6 per cent of their original genetic wolf heritage, yet the small difference of just 0.4 per cent is very significant. Toni eloquently describes why even socialised wolves will never become domesticated like the dogs with whom we share our homes and lives. Their independence, particular problem-solving skills, and sheer physical power remind us that, although distantly related, wolves are a separate species to dogs, with very different needs when humans come into their environment.

This book made me laugh out loud numerous times. It also moved me to tears. Tender and compelling, written from a perspective of profound understanding, affection and respect, Toni's stories remind us of how much we can learn from the creatures with whom we share our beautiful planet.

Lisa Tenzin-Dolma
Author
The Dog Welfare Alliance
The International School for
Canine Practitioners

Introduction

I am part of a very excusive club; one to which only a handful of people worldwide belong. I have had the privilege of working with socialised wolves. Not the scared, demoralised wolves you see in some zoos, who pace back and forth, back and forth, but those who have been reared to be comfortable around people. Wolves you can touch, and who allow you into their world: wolves who welcome interaction between our two species. A relationship between human and wild canine is unique. Their physical and mental strength is far superior to ours, making us the more vulnerable, yet, somehow, it's possible to build a friendship, gain their trust, and walk among them.

For ten years I spent time with these beautiful, powerful creatures at the UK Wolf Conservation Trust (UKWCT, or Trust) in Berkshire, England. Over the years my involvement increased as I progressed through the training to eventually become one of the senior handlers and head of the wolf welfare team. I also took on the role of education officer for four years, teaching school, college and university students about wolf behaviour. As I spent more time – not only with the wolves, but also experienced people like senior handler John Denness, and founder of the Trust Roger Palmer – I was guided and taught the difference between wolves and dogs. I spent hours with John and Roger, working with and watching the wolves, observing them through their seasonal changes, and at their daily activity peak in the crepuscular light of dawn and dusk. I nursed them when they were sick or injured, raised abandoned neo-natal cubs, and lazed around with the adults in their enclosures. I learned to howl with them, interpret their complex communication skills, and watch how they interacted with each other and us. No part of the wolves' lives was off-limits to me; I lived, breathed, and slept wolves.

The unique insight I gained from my time at the UKWCT has enhanced my knowledge of and work with domestic dogs. I can speak with authority about the similarities and differences between wolves and dogs, dispelling myths surrounding dog training, and help others successfully interact with their canine companions.

Writing this book was a selfish act, in part. I experienced so many heart-stopping, stunning moments with the eleven wolves I worked with I don't ever want to forget them, or let time dim their memory. Included in this memoir are some of the worst times – like nursing Alba through a spinal injury, sustained in a freak accident – and also some of the very best. Simple things like the wolves howling to call me back when I drove away at the end of the day; laying in the sun with wolves sleeping a few feet from me; a wolf cub climbing into my lap to eat an apple. These are the moments in your life that define you, provide you with serenity, and sustain you through the hard times.

These memories are very, very special, and I want to share them with you.

Toni Shelbourne

The characters in this story

Wolves, like people, have distinct personalities. They exhibit fears and phobias, likes and dislikes; get on with some people or fellow wolves but not others. To work with them it's necessary to treat them as individuals, and accommodate their idiosyncrasies. Building a relationship based on mutual trust and respect is essential for your safety and wellbeing, so please don't be hoodwinked by my tales: make no mistake, wolves are powerful animals, possessing a reputed psi of 1500lb (680kg) in their jaws, a bite from which can cause serious damage.

Working closely with socialised wolves should not be treated lightly – it's a bit like the part in a wedding ceremony where the priest tells you that marriage should be taken seriously – but, with the right skills, attitude, and knowledge you can enter the fascinating world of the captive, socialised wolf. Wild creatures, still (no wolf can be domesticated but they can be socialised), they have learnt not to be frightened of people.

So, let me introduce you to the characters in my story.

Essentially, the wolves lived in four groups, comprising Kenai and Kodiak, the North Americans; Duma and Dakota, two more North American-type wolves; the Europeans, Alba, Lunca, and Latea, and the Canadians, Torak, Mosi, Mai, and Mika.

Dakota

Dakota was the first wolf I ever touched, and for that reason alone remains a firm favourite. Born at Woburn Safari Park on 12 May 1998, she came to the UK Wolf Conservation Trust with her sister, Duma, when they were just a few days old, considered surplus to requirements by the Park (not unusual in those days when contraception for zoo animals wasn't as advanced as it is now).

Dakota was almost three when I first met her, and the omega wolf (as in a lower-ranking individual). To people, though, she was a wolf of two halves: very loving and affectionate, but also the most mischievous madam, who wasn't above testing out new handlers by either bouncing all over them or chewing their shoelaces. She could be walking along on a lead (all of the wolves were used to a collar and lead, and were walked regularly) quite peacefully, when a sudden glint in her eye meant she was up to no good. Dakota it invariably was who grabbed someone's clothing, or a bag left just within reach: you couldn't take your eyes off her for a second.

I remember one day, when on a training walk with new handlers, she was showing a lot of interest in another handler's brand new raincoat. I questioned if there was any food in the pocket: "No" was the answer. The next thing, Dakota had grabbed the material in her jaws and begun to tug, and when a wolf locks on to something and tugs, all you can do is try not to be pulled over: it's not like you can teach them a 'leave' command, or prise open their jaws, as you'd be bitten (and you wouldn't be able to open the jaws anyway, they're just too powerful).

The problem was resolved this time when the coat ripped in two, and it was only then that the trainee handler remembered she had wiped hotdog juice from her hand onto her coat!

Dakota can be described as obnoxious submissive. Dog trainers talk about active submission (actively seeking to show submission through muzzle-licking, pawing and crouching posture in front of another dog), and passive submission (rolling over in front of a more authoritative animal, waiting for punishment to be dished out or for the individual to walk away). Obnoxious submission is exactly what it says on the tin: an individual who is so over the top in their need to submit that they became a real pain in the neck, and Dakota used to do this to Duma, her sister, who eventually would either walk away or really tell her off with ritualised aggression. It drove Duma mad.

Being the omega among the wolf population at the UKWCT might mean one thing, but Dakota was still a wolf with attitude and ambition, who felt it only right that *she* had someone to pick on. And it didn't matter who – work experience students, teenage boys, new volunteers – she wasn't fussy. For this reason, she was walked only by people she trusted and respected: those she felt had authority over her could manage the situation when she got that glint in her eye by quietly saying her name, accompanied by a shake of the head, which would settle her. If she didn't respect you, heaven help your shoelaces!

Even though Dakota was a very well socialised wolf, her true instincts were never very far from the surface. To run away from unfamiliar or frightening situations is the first line of defence for any wolf, so although Dakota would be happy and unfazed at a busy event, a wheelbarrow out of place at home could be a big deal for her. She would refuse to walk past it, or wait until I did so in order to use me as a shield between her and this scary object. She was also wary of the metal traps and doors in her enclosure, especially when they rattled in the wind. Often, though, vocal reassurance and a belly rub would calm her.

Dakota loved to sunbathe, and would often lay out to bake for most of the day. In the summer sometimes she would be so comfortable she wouldn't even come in for her dinner, and you'd find her late at night, lying sleepy and content in her scrape (a bed that wolves make in the dirt). She'd let you stroke her, lay down next to her, or even use her as a pillow, but try to get her up and moving towards the kennel area where the wolves fed and slept and she'd become grumpy, so we usually left her there to enjoy the evening. Wolves don't need to eat every day, and will self-starve if they feel so inclined.

Photographic days at the UKWCT were a favourite with Dakota, who was always a bit of a poser (in her younger days, before she lost part of her tail and became ill with lymph node cancer, she was extremely photogenic). Dakota had her limits, even so, and after striking various poses (she seemed to have a stock of them, displayed in order: sideways look, sit, ears picked, howl), would let us know she was done with her adoring public. She always ended each session with a howl, after which she'd look at the handlers in a bored fashion, which definitely said "Right, time to go back and have a snooze now, please, I'm done toiling for the day."

Towards the end of her life Dakota was plagued by ill health, but although she became slower she never lost her spirit. She was gracious about and tolerant of all the treatments we administered, and she often had to stay on the yard for extended periods as she healed from various operations. Beating all the odds, Dakota lived with cancer for over three years, and it broke my heart the day we lost her.

In her lifetime she had appeared in countless TV programmes; eaten goodness knows how many shoelaces, and stolen everyone's heart. I can safely say there will never be another Dakota: she was unique. Kota, as I called her, and her sister, Duma, were possibly the best and most versatile socialised wolves in the world. I miss them every day.

Duma

Born alongside her sister, Dakota, at Woburn Safari Park on 12 May 1998, Duma was by far the most charismatic wolf anyone could hope to meet. She was special with a capital 'S.' Nicknamed the Queen of the Wolf Trust, for the whole of her life she was our 'go to' girl for every situation that required a wolf to do something extraordinary. She was the perfect ambassador wolf: calm, tolerant, amenable and adaptable. Whether she was interacting with a person in a wheelchair, comforting a sick child or wowing a presenter in front of a TV camera, Duma loved her work, becoming depressed and demanding if she wasn't able to do it.

One year I took her out of work for a couple of weeks as she was suffering from an open pyometra (a serious condition where pus collects in the uterus). She needed time to recover from this, but although she was ill, Duma was champing at the bit to get back to work. The satisfied look on her face when I finally deemed her fit enough to interact with the public again was hilarious, seeming to say "This is my job, and don't you ever forget it!"

She was also a born leader: authoritative without being overly aggressive, asserting just the right amount of control necessary to keep her sister in check. She'd often sort out an over-exuberant Dakota if she became too bouncy around handlers. We only had to wait for Duma to step in, growl, and display an assertive posture for the disruption to settle.

Among the handlers Duma had the nickname Miss Nippy Nicker, referring to her preferred form of greeting. Wolves love to muzzle-hold in acknowledgment of each other's presence, and would often do this to us as we entered the enclosure, jumping up to hold our faces. Not so Duma, however: she always wanted to nibble your chin, and if we tried to prevent her, by pushing her away, she became upset, and she'd redouble her efforts, coming at you harder and faster. This wasn't the nicest of experiences, but at least you knew it was only in greeting. It was an easy matter to distract her with a belly rub, and often she'd fling herself down at your feet and roll over for the full experience. If she happened to be standing and you found the right spot, she'd splay her hind legs further to allow better access, a gesture that's known as an inguinal presentation; a normal posture when wolves extend a welcome.

Although Duma controlled her sibling well, and lived all her life with Dakota and later Kodiak, she hated Kenai, Mosi and Mai with a passion. Duma, together with Dakota, Lunca and Latea, had been part of the group that deposed Kenai as alpha female when the wolves lived together at a younger age. When Kenai and Kodiak went to live in a separate enclosure for Kenai's comfort and safety, walking Dakota and Duma past this was a challenge, as Kenai would run at the fence to agitate the girls, who would, in turn, be furiously lunging back at her at the end of their leashes. Accompanied by snapping jaws and fearsome growls, all three would be grabbing at fence posts, and even, on occasion, attempting to re-direct their aggression to the handler's leg! It happened to me, once, when, in frustration, Duma turned, took my knee in her jaws, and began to exert pressure. A sharp word brought her to her senses, thankfully, and she let go, flashing me what seemed to be a gullty glance. Seconds later, however, she was lunging at the fence line with me holding on for dear life. Until you've been on the other end of a wolf in full flight, you will never have felt power like it: large dog or not, there's simply no comparison!

I firmly believe that Duma would have made a perfect mother, and regret that she never had the opportunity. She longed for a mate who adored her. Later in life, after Kenai passed away, Kodiak moved in with Duma and Dakota, and, during breeding season, Duma did everything she could to encourage Kodiak to court her, but he was clueless in that department, as he had been all his life. Duma would muzzle-lick, paw, present her rear end and avert her tail, to indicate she was ready for him to mount her, and some days she'd even try backing up to present herself to him. Poor Duma did everything possible apart from plant a neon sign above her head, but Kodiak either ignored her or ran away. Occasionally, when being 'kissed' he'd even shut his eyes and turn away from Duma (who'd be snuggling up to him) so that they'd end up walking in circles. Sporadically, he would attempt to mount her but it was always from the wrong end. You could see Duma thinking, "What a loser!"

Duma, you were the ultimate ambassador, converting so many people to wolf lovers, and helping us raise thousands of pounds for your wild cousins. I hope you run wild and free in wolfie heaven ... and have found a worthy mate at last.

Kodiak

Kodiak was the only wolf I didn't go into the enclosure with. By the time I joined UKWCT he had been retired from working with the public due to becoming intolerant of new people (like many males he was very dominant at maturity). This isn't unusual for socialised males, and all three of those at UKWCT – Alba, Kodiak, and Torak – had retired from public walks by the time they were three or four years of age, although Alba and Torak remained close to a number of handlers.

Kodiak only really loved one person and that was Colin Thorne, one of UKWTC's senior handlers. They met when Kodiak was young, and Colin continued to work with Kodiak his whole life, the only one who went into his enclosure and walked him. Towards the end of Kodiak's life I could walk his sister, Kenai, together with him, although Colin kept a strict eye on Kody to ensure I was safe.

Kody was big, bold, and didn't take any prisoners, and my heart was always in my mouth when Colin went into his enclosure to demonstrate to the public that he was just a big softy, really. Colin would playfully tug his tail and laugh at him whilst Kody strutted around growling and posturing. Kodiak, I have to say, never bit Colin.

Kodiak was another Woburn Safari Park wolf, and came to UKWCT in 1994, when just a few days old, to be hand-reared by Roger Palmer as part of his private collection. Kody and Kenai it was who were the first wolf ambassadors when Roger founded the UK Wolf Conservation Trust in 1995.

Kodiak was never the smartest wolf on the block, and spent his life being henpecked by one female or another. Kenai was always telling him off, and later, when housed with Duma and Dakota, he was constantly driven away from his dinner until we took pity on him and fed him separately. He had the best aggressive pose, though, of all the wolves, and could really put on a display for the public. He also used to poo just inside the enclosure gate as if to say "This is my territory – keep out."

In later life Kodiak suffered from arthritis and sebaceous cysts, which frequently burst and required cleaning. Roger had trained Kodiak well, and he would come on to the yard that led to the kennel blocks, to stand pressed up against the fence so that I could treat him via access holes we'd cut in the fence. You had to keep a close eye on him, though, as he could suddenly turn to grab you, and if he got hold of a finger or clothing, he took his time letting go. He was a stunning, majestic beast nonetheless, and never failed to impress with his posturing and vocal displays.

I spent a lot of time caring for Kodiak toward the end of his life, and, although he tolerated me, I can never claim he adored me as some of the other wolves did. He really only had one love – Colin – but this didn't stop me doting on him. He was a character, and very much a part of UKWCT's history. When he did finally pass away, at the grand age of sixteen years, he was one of the oldest wolves in captivity. Although his sight had diminished, and his health in the last few months had been poor, Kodiak was the most impressive wolf I have ever had the pleasure of knowing.

Kenai

The first thing that springs to mind

when I think of Kenai is that she was a madam, and not always in the nicest sense of the word: 'bully' is another description that fits. That's not to say that I didn't love Kenai, of course, because, mostly, she only acted this way toward other wolves.

As the oldest female wolf on site, Kenai had been thrust into a matriarchal role for which she was ill-prepared. With four other younger females to keep in check, and a lifelong mate to jealously guard against young upstarts, you can understand why this might be. A small, intelligent wolf, Kenai could defend her corner, and proved, for a time, that brawn was preferable to brains when it came to controlling a large pack of younger, smaller wolves ... until they grew up, of course, when they no longer tolerated her bullying behaviour.

The first time I came face-to-face with Kenai was the day I was invited to meet the wolves, and she and the other four females erupted onto the hardstanding to greet me. Kenai didn't really care about me being there: she was more interested in putting the others in their place, snarling and snapping at them around my feet. A short time later, she and her brother, Kodiak, were permanently removed from the main group as her bullying had become intolerable for the other wolves. Led by Dakota, the pack ousted her, and a nasty fight ensued, which handlers had to step in and put a stop to before serious injuries were sustained.

After this, Kenai lived out her life with her Kodiak in a separate enclosure, although this separation didn't prevent her taking every opportunity to wind up the other girls, and walking any of them past her enclosure – as they flew back and forth growling at each other and trying to grab the fence – made taking control somewhat difficult.

In truth, I had little to do with Kenai when I first joined the Trust. She had already retired, along with Kodiak, whose intolerance of new handlers meant we rarely got to interact with Kenai, other than having a fuss through the enclosure fence.

Kenai was astoundingly intelligent, even for a wolf. She was the reason why we had to put a safety clip on the metal check chain collars we walked them on, as she would run to the end of the six-foot lead, stop abruptly, and when the collar became loose around her neck, duck out of it. Rumour has it she even climbed out of a temporary enclosure when maintenance was being carried out on the main enclosure, and once, when out walking with the public, she went into the pond and caught a fish! She could outsmart the smartest person.

Toward the end of her life, we did establish a working relationship, and I felt she finally recognised me as an ally, which meant I felt comfortable to go in with her on my own and give her some well-earned, special attention. As with all of the wolves in my care, I nursed Kenai until her death (in the last few years, she suffered from a reccurring cancerous growth). As with all of the wolves, her level of tolerance when it came to receiving treatment – and what she allowed me to do in this respect without losing trust – just astounded me. We may not have had the best of relationships in the beginning, but toward the end Kenai and I came to an understanding, and she accepted me into the very exclusive club of handlers who worked with her. If Duma was Queen of the UKWCT, Kenai was grandmother, and her contribution should not be under-estimated. She was, after all, a founding wolf of the Trust.

MIKA

That same sad June of 2006 when Kenai passed away, we lost another little life – Mika – who was one day short of eight weeks of age. Her's and her sister's stories are covered elsewhere in the book, but what I can say here about Mika is that she was a joy. If she had been well, who might have emerged as top female in their little pack would have been a close call between her and her sister, Mai, even though Mika was half Mai's size. Each time Mai began posturing or playing too roughly with Mika, she would fight right back and tell off her bigger sibling: I have some wonderful video footage of the interactions between the two of them.

Mika means 'wise little raccoon,' and the name suited her well. Rejected at birth, she did not receive the vital first milk – colostrum – from her mother, so necessary to provide immunity and the best start in life. Over several weeks we noticed that she was struggling, and not thriving as well as the others, though, boy, did she have a big and feisty personality!

It was discovered that Mika had cataracts, and a bone density problem that meant her bones broke easily. The tibia of her hind leg was broken one day simply by one of her sisters standing on her. We thought about applying a cast, and isolating her from the others, but she cried endlessly when we tried this. After an x-ray and resultant bone density deficiency diagnosis, we knew it simply wasn't viable; she could break another bone at any time.

On the back of Kenai dying, June 2006 was a terrible month, especially for me and John Denness, senior handler and wolf welfare officer, who had to make the horrid decision about wolf wellbeing, which included end-of-life management.

My fondest memories of Mika are of the day my good friend, Lorna, left the Trust. Lorna worked in the office but, as a big animal lover, spent as much time with the cubs as she could. It was a lovely hot afternoon, and we all gathered on the picnic benches outside the main enclosure to hand out gifts and toast her farewell. The cubs were just at an age where they had begun to leave the den and explore the surrounding area, so we let them out of the building they were housed in, and they mostly followed along at our heels.

All, that is, apart from Mika, who went in a completely different direction and managed to find and go through the only gap under the outer perimeter fence, and into the garden beyond. After a mad dash we caught up with her, just before she fell into the pond there. That day she played with plant pots, carrying them around like prizes, chewed on plants, play-fought with her siblings, and eventually climbed on to my lap for a snooze in the sun after demolishing a feed.

Mika may not have lived for very long, and she looked raggedy and small, but that little cub sure did steal a lot of hearts. I'll always remember her by my nickname for her – Tiny – as that's how I identified her before she was named. She and I both had a difficult start in life, but had learned to hold our own. She reminded me of myself, and I admired her spirit.

Mai

In the Native American Navajo language, Mai (pronounced My) means coyote, and she was the biggest, most gregarious of the three female siblings I raised (also Mosi and Mika). After picking them up from Dartmoor Wildlife Park on 8 May 2006, I called Mai 'Bib' at first, simply because she had a white patch on her chest that resembled a child's bib.

Born on 27 April 2006, Mai was twelve days old when we first met; her eyes just opening. As a cub she was big and bold; as an adult the gentlest, sweetest wolf around people but nervous of the world in general. I often described her as having a split personality – fairly common in wolf society – always ready to put Mosi in her place and take full possession of Torak, who she loved, but more than willing to let Mosi go first in new and scary situations.

Mai loved people, and often chose to hang out with her 'people pack' rather than her wolf pack, remaining long after the others had become bored and wandered off. She would often rest her head against my chest and lean into me for a fuss, or chin-rub my head if she was on the greeting platform and at head level. As an adult she appeared to dislike cold, wet mornings, and would often go back inside the kennel when let out. She'd happily monopolise me for a cuddle, growling at others if they came near. Perhaps, after all, it wasn't the weather that drove her back into the kennel where I'd be working, but a need for interaction with one of her favourite people.

Mai was a bit of a foodie, and usually came on to the yard with no trouble once she was used to the routine. In fact, she could get so enthusiastic, she would sometimes forget herself, and bounce up and grab clothes – or even hair – in excited anticipation. Many's the time I had to remind her to be gentle, and not swing off my pony tail! She didn't mean to hurt, she just couldn't contain herself when she was younger.

Her love of treats made Mai ideal for photo days and student research projects, which often involved novel smells or food. The only problem was she preferred to hang around the fence line, asking for a fuss instead of showing off to the public up on the mound. We had to resist stroking her to discourage this behaviour, but you could see she didn't really understand why her adoring public wasn't forthcoming with caresses.

From a very young age Mai threw her weight around with the other wolves, claiming the breeding female status and teaming up with

Torak. I'm not sure Torak cared who he was paired with, but Mai adored him, and would aggressively warn off Mosi, especially in breeding season when the hormones were flying. I can picture Mai, a few weeks old, riding up on the shoulders of the other cubs, growling, posturing, and generally fighting her corner – behaviour which continued for the best part of her early years until Mosi finally seized power when they were about three or four years old. Mai, on the whole, was a good leader, though: not as natural as Duma but not a bully like Latea and Mosi.

As youngsters I taught the cubs to be comfortable around water, and, most importantly, the containers we kept the water in. Mai loved water in all its forms, be it in the hot weather to cool off or as ice that she could break up and play with. If they weren't in the enclosure that held the pond we provided water troughs for the wolves to play in. All three – Torak, Mosi, and Mai – could be seen jumping in and out; kicking up the water to cool their bellies; laying in it on hot days. Mai often appeared to try and drown Mosi, jumping on her shoulders and wrestling, trying to push her down, all in the name of fun!

Blackberries were Mai's favourite fruit. She'd pick them off the bush with her incisors, though if a hander picked them for her, she'd willingly take these as well. Hotdog-flavoured ice lollies were another must-have on a hot day to cool off: she went mad for them.

Mai's black coat used to be rare in the wild but is becoming more common. It's a trend that appears to have started with Canadian wolves, and Mai is what we used to call a Mackenzie River Valley Wolf (*Canis lupus occidentalis*), now more commonly known as a North-Western Canadian Wolf. Scientists think that the black-coated gene originates from cross-breeding in the wild with domestic dogs, which can happen if a female is without a mate and a wolf and a dog meet at an opportune time. She always had a distinctive white mark on her chest, as previously noted, but, over the years, Mai's coat turned grey, and she resembled her mother, Lizzie, much more. If you didn't know her, or understand coat types, you'd think she was old before her time, but no: greying can be a natural process in early years as well as in old age.

All wolves are naturally destructive, with a strong instinct to chew and explore with their powerful teeth and jaws, but Mai could take this to extremes. Each morning I wondered what I might find: another young tree chewed to a stump, or a plank ripped off a day kennel and used as an object of play. And hunting came naturally to Mai: I saw her first pounce on mice at barely ten weeks of age – and it wasn't long before she was catching them. Pheasants were her other favourite, and help from her pack mates meant that a fair few met a grim end whilst trying to escape the enclosure.

A couple of things that Mai couldn't abide were umbrellas, and the sound of people walking over gravel – both freaked her out – and from an early age she was suspicious of many things that the others, especially Torak, took in their stride.

She may have been bombproof with people, but the world around her, outside of the enclosure, could be a challenge for Mai. Wheelbarrows left by the path; a little wooden bridge out on one of her walks; the trailer: all proved difficult for Mai to cope with, and just how difficult depended greatly on who was handling her. Vocal reassurance made matters worse, and, although it sounds callous to those unfamiliar with canine psychology, if you ignored her and remained confident, she'd quickly move on, reassured that you were unconcerned. It was partly because of Mai's behaviour on walks in the woods that wolves were then only walked at home on land owned by the UKWCT. The poor girl would get herself in a right pickle, and it wasn't fair on her. She much preferred being on familiar territory.

Because of her love for and confidence around people, I began training Mai to work with children. I used to run the children's walks and wolf keeper days for young Trust members, and although Duma was the star in this respect, she was growing older, and I could see a need to replace our Queen one day. Mai took to the role well, and became a firm favourite with our younger visitors. Always gentle, always patient, she would stand for ages letting the children approach, one by one, and stroke her belly. Torak and Mosi never really enjoyed this part of the walk, but Mai – true to form – accepted most people as friends, with very few exceptions.

I loved Mai: she wore her heart on her sleeve, and was never nasty or malicious. Such a gentle, honest soul.

Mosi

In contrast to her sister, Mai, Mosi was the rebel without a cause. A minx from the moment she could 'talk' (growl and snarl!) – cheeky, mischievous, always pushing her luck – Mosi was a perfect troublemaker! Of course, that's not to say she was nasty, or that I didn't love her. She had spunk and spirit in equal measure, and could stir up trouble in a heartbeat.

As a cub, she'd always be hanging off your clothes or shoes, and always had to have the last word. She showed little initiative, preferring, instead, to follow others to hidden food, rather than sniffing it out for herself, and try to steal or beg it from them. She'd jump off anything, without the slightest clue about where or in what she might land, and she is the first and only wolf I've seen blow bubbles underwater through her nose!

Mosi kept you on your toes. She learnt how to take earrings out of ears; suss out the weakest link in every group of visitors we took into the enclosure, and, in later years, pushed Mai to the limit with her ambition to be the breeding female, which she eventually achieved. In essence, Mosi was the 'go to' girl for fun with a capital F, and anything that got in the way of goofing about was frowned upon. Even feedtime was occasionally ignored in favour of a game with Torak out in the enclosure: who needed food when there was a whole host of smells, and prey to chase and catch in the enclosure?

Mosi had a wicked sense of humour, and her antics always made me laugh, even when I was trying to extricate a visitor's clothes from her mouth, or explain that Mosi growling during a meet and greet (the public was invited to stroke the wolves) was 'just her way,' and not to be taken seriously. I often demonstrated this by kissing her on the head when she was in full grumble mode. I suspect this made many people very nervous: it probably wasn't the most sensible thing to do, after all, but Mosi and I had an understanding. We respected one other, and knew the limitations of what was acceptable behaviour around each other.

As a lower-ranking cub, the pack would have regarded Mosi as expendable. Subordinate wolves, or sub-adults as we call them, often go first to investigate new objects in the environment, or are sent forward whilst higher-ranking individuals or the parents hang back.

As a youngster, Mosi appeared not to have a care in the world: chasing butterflies, jumping off rocks, walking over piles of logs, and leaping into water. Nothing seemed to faze her much. Mosi means 'cat' in the Navajo language, and I swear she used up many of her nine lives in her first year. She wasn't very bright, though, and would often fall for Torak's and Mai's cunning, especially when they wanted her food. Using play to distract her, the pair would then run off with her booty: easily lost because Mosi didn't have the brains to work out what they were up to.

Mosi loved water, and was constantly playing with and emptying the water buckets – partly my fault as I trained the cubs to be comfortable with metallic-sounding objects. They would often put their heads inside the empty bucket as I tapped the outside, turning their ears this way and that to locate the sound. Very cute … until they began running off with the bucket!

Out of all the wolves the Canadians had the loudest, scrappiest howl, more like yowling than the melodic notes of the North American and European wolves. Mosi's was the worst, a screech that went from high to low octaves, and back again. Howling would always trigger a mass muzzle-holding and play session, which Mosi would always be in the thick of, jumping up, grabbing her pack mates, and me, if I happened to be in the way. I remember once being pulled in opposite directions at the end of one enthusiastic early morning howl session: Torak had my sleeve and Mosi my trouser leg. Lots of sounds would set them off, from church bells, to howling by other packs.

Mosi participated in many student research projects, and played her role of subordinate beautifully. In the wild she would surely have been dead before her first birthday, because many of the things she should have been wary of were just toys to her.

This was illustrated beautifully by an experiment with new kinds of stock protection fencing. Farmers in Europe – and now in the States – use a simple fencing system called flagery: essentially, red strips of cloth hanging off a wire fence. The student doing the research experimented with other items – such as CD – all of which Mosi just wanted to paw at and, if possible, grab. Not the reaction we were looking for. She was just too curious: potentially disastrous for a wild wolf.

Mosi was always a challenge but, boy, did she make things interesting; her zest for life was infectious.

Torak

Torak was born the same year as Mosi, Mai, and Mika but at the Anglian Wolf Society. A tall but lean wolf, he was the result of a mating between a European male and a North American female. This made him a little odd-looking, as his body seemed to want to contain the builds and coat colours of the two separate subspecies. There are few categorised species of wolves but many subspecies. In the wild, of course, Torak would not have existed, as his parents came from different continents.

Long in the body and very tall in the leg, without his impressive winter coat he was very lean-looking. As a cub Torak was much bigger than the girls, and we often had to separate them for short periods to give the smaller cubs a rest, as Torak just wanted to play the whole time. He was strong and adventurous, quickly learning how to climb up on the straw bales which made up their den (usually if he wasn't fed first and wanted his bottle).

From a young age objects and environments didn't worry him – he was bombproof – but after a particular scare, he became wary of some people, especially men. Shy and jumpy, then, he often elected to remain on the periphery when his least favourite people entered the enclosure. In time, this made him tricky to catch and collar for walks, but once out, he loved exploring. He was never truly comfortable doing meet and greet, however, and we didn't push him.

With his main handlers, Torak was very comfortable and affectionate, often jumping up to muzzle-hold us. Unfortunately, as it was not unusual for him to be carrying a scrap of meat or fur in his mouth when he did so, this was not for the faint-hearted. He very soon grew bigger than me, and, with his paws on my shoulders, towered above me.

Although much bigger than the girls in his pack, Torak was a sensible boy and kept out of trouble: often seen heading out of the fray if the girls were having a tussle. That's not to say he didn't enjoy play-fighting, but if things got a little out of hand, he'd be off in the opposite direction. Nevertheless, if one or other wolf was bullying a sister – which mainly happened as they matured and hormones came to the fore in breeding season – he did not hesitate to step in to control the level of aggression. This is typical behaviour for captive male wolves, who will allow behaviour to go so far, probably because they want to mate. But push them too much and they will always put the girls in their place.

Torak was late to mature: it took him ages to cock his leg when urinating, and he never did this consistently, even when around two years old, when he did eventually begin to do so. Wolves mature very slowly, and don't attain full mental development until around three years of age.

Torak did have one pet hate and that was going into the kennel. He would need to be really hungry to allow himself to be shut in the narrow space, and, once the food had been consumed – in seconds – would scratch at the door or metal traps, making a huge racket. If you walked away from him, mostly he'd quieten down, and letting him out when he did this just made him worse. In the end we struggled to get him on to the hard standing even for food, and once he began losing weight we had to feed him in the enclosure rather than the kennel. He wouldn't come on to the yard if what he considered to be 'scary' men were nearby, either, but tease you by putting his head just inside the gate and then running off.

Torak loved fruit as well as meat and bones, and I remember him, as a cub, curling up in my lap to eat an apple. He could unpeel bananas, loved blackberries, and once fought me for possession of a pumpkin – I still have the scar to prove it!

He also disliked open days, when the Trust was packed with people, and would sneak off to the woods to hide out for the day. We'd constantly be asked where Torak was, but he would never show his face; not even if handlers entered the enclosure. We would have to put an extra water bucket out for him as he wouldn't even come in for a drink.

Torak put on an impressive display of hackles if he was riled by the girls; fought for food if need be; captured the hearts of many, but let only a chosen few enter his inner circle. I'm privileged to have been one of them.

Alba

Alba was the male of the European pack, and lived with his sisters, Lunca and Latea. This litter was born at the Trust on 3 May 1999, and is said to be the first born on British soil since we eradicated wolves from mainland UK, hundreds of years ago. Roger Palmer imported their parents and an aunt – Apollo, Athena, and Luna – from Europe, and they subsequently went to live at Wildwood in Kent.

To various handlers Alba was a wolf to fear. Domineering; sometimes unpredictable, he was an impressive, majestic animal who took no prisoners. I never understood why he accepted me into his inner circle and not others, and I never once felt threatened by him. Even at a very early stage in our relationship he would enthusiastically greet me by sometimes holding my entire head in his mouth!

I didn't meet him until he was about eighteen months old, and I was one of the last people he took to. Volunteers who joined after that time were held in distain, or targeted for an impressive display of posturing and growling. He could set a handler's heartbeat racing in a minute if he looked at you the wrong way, and had the ability to stare you down with hard, glazed-looking eyes. If you saw that look, watch out!

To me, though, he was just Alba, my mate, and we rubbed along together well. Although he was too strong for me to handle on a lead on public walks, I occasionally got to walk him double-leaded with one of the strapping male handlers that Alba liked. His power was amazing, and I'll never forget the sight of him dragging around sixteen stone guys when he got the wind up his tail, which often happened when he'd had a dip in the pond on walks, and had become frisky. I've no idea how they managed to hold onto the lead ...

I never really knew how Alba would be in the enclosure: some days he would want to play; others he simply rolled over for a belly rub. Either way he demanded respect and compliance – and he was bigger than me, after all!

Even though Alba had a luxuriously thick coat, which was very waterproof, he hated the rain. He would often lounge around in his comfy, warm bed in the kennel, refusing to get up when the pack was let out at about 6.30 each morning.

After being kicked out post-fuss, Alba would simply seek the shelter of one of the day kennels, usually within sight of the feed room. Alba loved his food, and would often steal the lion's share of treats (stuffed melons or meat-flavoured lollipops). Later on in life, after the devastating accident in which he fractured his neck, Alba's medication was concealed in sausages, and he was fed separately, as the girls would knock him out the way otherwise.

In some ways Alba's accident defined his life. I tend to remember him pre-accident: young, healthy, proud, strong-willed; loyal to those he liked and intimidating to those he didn't. He loved attention and being fussed, but only on his terms; when he wanted it. He enjoyed life to the full and was resolute in the face of adversity. I admired his spirit. Goofy at times, very affectionate to his friends, a good leader and keeper of the peace between his sisters, Alba never lost the respect of his siblings after his accident and ensuing disability. His strength of character and physical fitness really saw him through all his trials and tribulations. I can safely say I'm one of a very few people who have nursed an injured wolf; the list of those who did this without getting bitten is even shorter!

Alba was everything a wolf should be: proud, powerful, intelligent, slightly intimidating but tolerant and affectionate. I miss him greatly.

Lunca

Lunca led a double life in some respects: initially a pushy pup who never really backed down, and later a lower-ranking wolf who patiently accepted what life and other wolves threw at her. As a young teenager she managed to wrestle female leadership from both Kenai and Dakota, and nobody, wolf or handler, stood in Lunca's way. She was dominant, feisty – a real go-getter who loved Alba with a passion – and, although they were siblings, they remained lifelong mates.

Wolves in the wild rarely inbreed, but Alba and Lunca had no choice, although we took steps to prevent any offspring. In the early days, if any handlers were going to be bitten badly, you could guarantee either Alba or Lunca – or both – would be involved.

Lunca was impressive-looking, standing nearly as tall as her brother, with a heavy-set body, wide head, and beautiful colourings: powerful in both body and personality. You didn't mess with Lunca. She and I had a rocky start, and not until her old age did I feel she really trusted, or even liked, me that much. She tolerated me, mostly.

In truth, at first I struggled to build a relationship with her, probably because she frightened the hell out of me! She would be fine one minute, then do something to make me wary of her again. Don't get me wrong, she never bit or overtly threatened me, but Lunca could read your mood and emotional state from a mile off, and would act on it, too: disciplining her was not possible if you didn't have her respect. If Lunca wanted to hold your fist in her mouth, she did so, whilst you had to wait for her to get bored with this – and hope she didn't bite down.

It was not just people she could be overbearing with, either. Latea, her sister, was brutally tormented, especially in breeding season. Nobody was having Alba but Lunca, and that was that. Even after Latea deposed her some years later and meted out the same aggressive treatment, Lunca and Alba still mated in breeding season.

Lunca's other great passion was food, and could that wolf eat! After Alba's accident, we noticed she had begun to pile on the pounds, and became quite obese. But it wasn't until we observed them through the cracks in the kennel door (they were fed inside at night at that time), that we understood why. Alba, being unstable on his legs, post-accident, was slower, and was knocked aside as Lunca dashed to get the choice bits of meat, with Latea in hot pursuit, trying to discipline her. But whilst Latea was giving Lunca a telling off, hers and Alba's dinner ended up inside her sister. Despite changing the feed routine, Lunca never really returned to her former slim build, mostly because others felt sorry for her and gave her extra food, even when I was trying my hardest to keep her on a strict diet.

Lunca's dislikes included noisy maintenance work on-site, big noisy vehicles, and hot air balloons, all of which sent her running for cover, barking as she went. We used to take the pack to Newbury Agricultural Show every year, a huge event where countless air balloons would take off at some point in the late afternoon. We had to ensure that someone she trusted was in the temporary travelling enclosure with her then, for fear she'd try and jump out, with the attendant risk of hurting herself.

In later life Lunca became very stoical, apparently accepting of whatever life, or Latea, threw at her. She had dignity, courage, passion, and – even with failing eyesight – a zest for life. Most of all, though, she adored Alba.

In her dotage, she was a far cry from her former self, though we loved her, and inwardly thanked our lucky stars she had mellowed from the difficult-to-manage character she was as a youngster. Lunca and I finally grew comfortable with each other. She would let me pick burrs from her coat, and groom out her yearly moult. She taught me a lot, but mainly never to take for granted my relationship with any of the wolves.

Latea

Awww, little Latea: what a character:

small but feisty; shy but trusting. I miss that wolf. I spent a lot of time handling her on walks with the public, as she needed a calm and confident approach as a youngster. I remember she would not want to greet the public in the line-up (all of the wolves had an opportunity to sniff the walkers before setting off), and would hang back and shy away. I could sometimes get her to walk past the line-up, or we would just wait and walk behind the group for a while, until she felt ready to walk among the public, sniffing hands as she weaved through the bodies.

As a youngster, Latea was fun, mischievous, a brilliant initiator of play sessions; cunning in getting what she wanted, though picked on a lot by her sister, Lunca, especially in breeding season (which, ironically, Latea did to Lunca a few years later, after she deposed her as top ranking female). Latea was not a good leader like Duma, and constantly harassed Lunca, keeping her down in case her sister – who was bigger – should try and regain the upper hand. Latea would try and pick on Alba, too, or use him to punish Lunca. Alba didn't take kindly to that: Lunca, in his eyes, needed protecting at all costs.

Breeding season was exhausting for all of them but more so for Alba, particularly after his accident, and gave me a few sleepless nights, too, trying to work out the best course of action to restore harmony within the group.

Latea loved water, and was always in it or playing with it if the opportunity arose, be it water trough or pond, in hot weather to cool off, or using the ice as a play object to interact with in the cold. She even attempted to drown her sister in it on a regular basis. Watch out when she got out and shook, though, as she'd usually choose to do this when standing right next to you, then walk off with her usual wolfie grin.

I'll say one thing for Latea; she had a sense of humour. She loved to play and could leap vertically into the air or to the side like a kitten. It was difficult to get her in to the kennel for her dinner some nights, she was too busy playing chase or hunting mice. She was known for wanting one last game before bed, dragging the others back out into the enclosure by their scruffs. She would even try to get me to play by running up to me and patting my leg, as if to say "You're it," then running off again at full pelt.

Always affectionate with handlers, Latea would stay around me for ages in the enclosure, and when up on the greeting platform – which meant she was at head height (or higher, in my case) – would rub her chin all over my head and snuffle my ears.

None of the Euros, as we called this pack, was easy to work with, but Latea was probably the safest. She had an endearing habit of running up to the enclosure fence, wagging her tail, and doing her version of a wolfie grin, and also sometimes bring you an object that she would parade up and down with. And it didn't matter if you were friend or stranger, you'd still get this greeting.

In her later years, Latea became more used to the public, and would greet walkers with enthusiasm – albeit a quick 'hello' and then off at a hundred miles an hour exploring, rolling in novel smells, or trying to chase pheasants. In some ways, the Euros were a handler's nightmare. A real baptism of fire, if you could handle a Euro, any other wolf was a doddle.

Latea's joie de vivre was infectious. She did everything at full speed, and, when being playful and sweet, was a joy to watch. She didn't, however, like to share. We would often give the wolves novel objects to play with, such as scented, straw-stuffed hessian sacks, or stuffed melons. Latea would grab hers and run off a short distance; not far, though, in case there was an opportunity to steal Lunca's and have two. I don't remember her stealing Alba's, ever: I don't think she dared try.

It was all or nothing with Latea ... and that is exactly how she died. One day full of life, then dead the next. She was found curled up in the straw in the kennel one morning, after apparently dying in her sleep. The vet thought she'd probably had an aneurysm. And that was the end of our sweet girl: the wolf with a double life.

The people

You may have noticed that I've made little mention of people, and that's because this is the wolves' story, based on my memories; personal to me. That's not to say, of course, that I did all of these amazing things on my own. There was always a dedicated team of volunteers present, working hard to give the wolves the best life they could possibly have in captivity. I made many friends, some of whom I regard as lifelong buddies. I wasn't the only one daft enough to be found at the Trust at the crack of dawn, or in the middle of the night, caring for these wolfie characters 24/7.

The facilities at UKWCT are the envy of many a wolf keeper and handler around the world, and include big enclosures and many enrichment features, all thanks to the vision, effort and financial backing of one man – Roger Palmer – who founded the Trust, though, sadly, is no longer with us. Roger had developed a brain tumour before I knew him, which robbed him of his short term memory, and meant he could be brash with people. It took him months to learn my name but, once he did, I often wished he would forget it again, as he was always hollering at me. Despite this, for some reason Roger liked me, and quickly involved yours truly in the day-to-day running of UKWCT. I used to run the seminars, which automatically gave me a place on the executive council.

Roger was always late, and would then rush around shouting orders to"Collar-up the wolves," "Load the trailer," or whatever. He was also a terrible motorist, driving way above the speed limit, regardless of whether or not the wolf trailer was attached. I learnt a lot from him, though, and liked and respected him. He knew wolves, and he trained them well: Kody and Kenai were a dream to put to bed, for example, and there were never problems with them not using the traps or hanging about for a game or a fuss. Straight in, no fuss, food and bed. Wolves not trained by Roger were never as well-behaved.

John Denness was the other main influence in my wolf life: a senior handler and head of wolf welfare and safety when I joined the Trust. All of the roles he held were carried out with vigour. Describing John in the context of a wolf pack, would make him the ultimate alpha male, as he was big, dominant, proud, arrogant; could make you cry with laughter or from a telling off. Like Roger, John knew wolves, and his favourite saying was "Toni, they are not dogs," either shouted at me or with an outrush of frustrated breath, usually as a result of my suggesting a course of action to do with the health of one of our charges.

The wolves were his babies, especially the Euros, and he and Alba where inseparable. Over the years, John and I came to an understanding. As long as I knew he was the boss, he would listen to me. I would often slip an idea into his head and let it swim around for a while. If he thought it was a good one, it would come out of his mouth eventually, as if it was his idea, and I would nod and say "Good idea, let's do that." John may have had a wolf's strength, but I could use wolf cunning when required. John was like Marmite: you either loved him or you hated him; there was nothing in-between. I grew to respect him, and, in turn, he eventually came to rely on me to offer the best health advice, and take the lead with caring for the wolves (although I had to go through the motions of obtaining his permission, for appearances' sake). He taught me wolf behaviour; I taught him how to look after their physical wellbeing, and in this respect, it helped that he was into alternative medicine, as I was able to get homeopathic vet and friend Nick Thompson to treat the wolves, which had amazing results.

John, like his beloved Latea, died suddenly, and not long after his wolfie soulmate, Alba. I hope they found each other again in the afterlife.

I will always be grateful to John and Roger. They were kind enough to share their knowledge with me, and, at the same time, generous enough to allow me to teach them a thing or two. I owe them both a lot, and I hope they know how much I appreciated it, even if I never got to tell them.

And now that you have met the characters, let the story begin ...

Love at first sight

The phone rang. A lady introduced herself as Gerry, a wolf handler for the UK Wolf Conservation Trust. She'd heard I practiced Tellington TTouch training, and asked if I'd be willing to come and "Do some with the wolves." Gerry had seen and read a little about this method of animal training, and tried it on the wolves herself, but was interested to see what it would be like if "someone who knew what they were doing" did it.

Tellington TTouch, amongst other things, can calm an animal and help them cope in stressful situations. Simple body work which uses pressured touch releases feel-good endorphins like serotonin and dopamine. The method also affects brainwaves, enabling people and animals to function calmly, precluding the fight or flight instinct, and facilitating rational thought. Wolves live by their instinct, and can be very suspicious and flighty at times: this is, after all, what keeps them alive in the wild. Gerry thought TTouch would help the wolves deal with unusual situations, such as introductions to strangers, or when on display at shows. She invited me to meet them.

Consequently, one sunny afternoon in March 2001, I arrived at the Trust in my car, and five curious wolves ran up to the fence, tails wagging, and giving me their wolfie grins (in those days, there were no safety barriers). Gerry and Paul, her partner and also a handler, invited me to hold my fist out so the wolves could sniff and lick the back of my hand.

It was love at first sight – on my part, at least – and I was totally awe-struck: all I could do was stare at these astonishing creatures. Previously, I had only ever seen zoo wolves, who hide or pace, but the Trust wolves chose to approach and invited attention. I marvelled at their beauty and obvious power. Each sought attention in a slightly different way, and vied for the prime position to receive a scratch through the fence from Gerry and Paul, pushing themselves into the contact: they loved it as much as we did. I'd never met a socialised wolf before, one raised by humans; they lacked the usual instinctive fear of mankind. Although I was warned not to lean on the fence, and take care as the wolves didn't know me, I was hooked. What beautiful creatures, and what an honour to be so close to them.

That meeting through the fence would have been enough for me, but what came next topped this first experience a hundredfold. Gerry asked if I'd like to work with one of the wolves, doing some of the Tellington TTouch body work. I agreed instantly, appearing, I think, more confident than I actually felt, and was taken on to a concrete area between the wolf enclosure and a block of kennels. The plan was to ask one wolf on to the yard for me to work with. In the event, all five burst through the gate, and I was suddenly the centre of attention for five, fully grown wolves. Luckily, I don't scare easily, as one was jumping up to sniff my face, with another growling and nipping at a third creature at my feet. Essentially, I was pinned up against the fence getting a full-on greeting – heaven! I hadn't completely lost all common sense, though, and slowly, without turning my back, slid along the fence to the nearest gate on to a second yard to extricate myself from the situation.

Eventually, Lunca, Latea, Duma and Kenai were secured in the main enclosure, and Dakota, a two-year-old North American wolf, remained, contemplating me on the hard standing. I went back through the gate, and instinctively began to approach her as I would a nervous dog, crouching, and looking away from her so as not to appear threatening. However, I was quickly instructed not to do any of these things, which a wolf would regard as a sign of submission, and you don't want to approach a wolf from the position of a lower-ranking individual.

Dakota sniffed me, eyed me up, and then, under instruction from Gerry, let me rub her belly. All the wolves at UKWCT are conditioned to allow strangers to stroke them in this area, a throwback to when they were cubs, and their mother would lick them here to stimulate urination or defecation, so a comforting and reassuring action. It certainly seemed to quieten Dakota, who stood with head lowered and eyes soft. She was free to move around if she pleased, unrestrained by lead or collar, with Gerry and Paul standing nearby. I introduced some of the body work while still in the tummy area: TTouches (specialised ways of moving the skin in circles; lifts and strokes) are warming and comforting, and not invasive. Dakota seemed to approve. I slowly worked my way from her stomach, and tried some variations of the not-so-stimulating TTouches I'd started with, all the while gauging her reactions (such as breathing rate) and facial changes, such as a worried or hard-eyed expression. If she seemed anxious I returned to the belly.

Very quickly I was able to touch most of her body as she voluntarily stayed beside me, remaining still; allowing the body work. I honestly could not tell you how long I stood on that yard, communing with Dakota. I completely lost myself in her, watching her expressions, reading her body language, wondering at her tolerance of a complete stranger touching her.

Her coat was thick, harsh on top with long guard hairs, and soft and warm closer to her skin with the insulating undercoat. Multiple hues spread throughout – greys, browns and reds; black and white. Her coat had little smell and was in beautiful condition. Her body was lean and powerful ... and those eyes! If a wolf looks you in the eye, it's like she or he is reaching into your soul. You can be accepted into their world or rejected out of hand in an instant, and if they should decide not to trust or like you, well, you will never win them round.

Eventually, Dakota wandered back out into the enclosure to be eagerly investigated by the other girls. I couldn't stop grinning; incapable of stringing together a sentence, I was so overawed. I wondered if Gerry would think I was a complete idiot.

It was only then I discovered that it was breeding season, when I was shown another, smaller enclosure that held two males – Kodiak, the older, less sociable wolf, and Alba, the young European; brother to Lunca and Latea. Back then neither boys had been vasectomised, and alternative birth control methods were limited, so the safest thing to do during breeding season was to separate the males and females. Not only had I met five unrestrained and very much in-my-face wolves for the first time, I'd done it during breeding season, when hormones were flying. What a brilliant day!

Chatting to Gerry whilst being shown around, I learned that all of the handlers were volunteers, and anyone could come and help out with cleaning the kennels where the wolves were housed overnight, assisting with public events, and so on. Once you had volunteered for a while you could apply to be assessed as a potential handler. So, guess what I did next ...?

GETTING TO KNOW THE WOLVES

Gerry was a whiz at recruiting people she thought would be good with the wolves, and with her support I began to visit the Trust on a regular basis to help out. New volunteers were encouraged to spend time around the wolves by cleaning out kennels, and supporting handlers on members' walks by supervising car parking, making tea, and talking to the public who have paid to walk with the wolves at the regular weekend events. A wolf's sense of smell is phenomenal, so it isn't necessary to physically interact for you to become familiar to them. When on the hard standing between the kennel area and enclosure, the wolves – if they chose to – could come and sniff you, and invite you to fuss them through the fence. I didn't approach them without a handler present, though, as the animals could – and occasionally did – grab clothes, and it was a struggle to retrieve your sleeve or whatever from them. Wolves have a bite power of around 1500lb (680kg) psi, and can lock their jaws, so clothing- or finger-grabbing was discouraged. The trick was to let them sniff or lick the back of your hand without putting any part of your anatomy through the fence; then, if they turned side-on and pressed up against the fence, give them a scratch. If you didn't know them well you were encouraged to touch them nearer their abdomen, in that safe area they had been accustomed to having strangers stroke. Once you knew them better, you could work your way out from there until you could touch them all over, with the head the last place to be touched: our wolves became very agitated if someone who hadn't gained their trust stroked them in this sensitive area. Wolves muzzle-hold each other in greeting, but also use muzzle-nips and holds to discipline, so aren't always comfortable with people touching them in this region.

Cleaning out their kennels was not always a pleasant business. The *canis lupis* digestive system works in a brilliant way. If fed a good, high protein diet of muscle meat, black, runny faeces result, and these smell foul. If lots of bones have been consumed, a wolf will eat a small amount of the prey animal's fur which coats the outside of their stools to protect the digestive system from fragments of bone passing through, and this also stinks. The wolves were fed inside the kennel, so you were never quite sure what gruesome sight might greet you after they'd eaten. Wolves are not renowned for leaving food, but often there would be blood up the walls if they tussled with each other over choice pieces of meat, or it could lay pooled on the floor.

As newbie volunteers we were invited to go on the private training walks with the wolves after the main event of the day was over. New and potential handlers need time to practise controlling the wolves before they can handle them in front of the public. The wolves are really strong, as well as fast and often mischievous, so skill is required to ensure the situation is safe and under control. You also need to know how to deal with whatever may develop, but most of all you want to earn the trust and respect of the wolves you handle, and this takes time, especially if you haven't worked with them when they were young cubs.

It can easily take six months to a year to build a relationship – longer with some individuals – and with others you never feel that they have accepted you: you're simply tolerated, or maybe even rejected. You don't want that to happen, of course, but the fact is that some people never make the grade as a handler due to a lack of rapport with the wolves. This isn't something you can fake, and there's nothing you can do to win them round if they simply don't like you. This may be because of a mistake you know you've made with them, but other times you can analyse the situation for days and never work out why they don't want you near them. I loved these walks, spending time with them, watching their body language and how they reacted to the environment. I began to get to know their individual characters, as their personalities really shone through.

Handler assessment day

A few months after I began helping out with cleaning duties, a handler assessment day was scheduled, and those who showed an interest in becoming a handler were invited on to the training programme if they showed potential.

I'd not met Roger Palmer, founder of the Trust, before, and discovered he was something of a character: definitely upper class, from 'old' money, forceful, often rude, and liked to shout a lot. As previously mentioned, some of these idiosyncrasies were due to a brain tumour he'd developed a few years previously, which also robbed him of his short term memory, so he never remembered people's names.

Roger and the other senior handlers didn't know I'd been on the yard with Dakota (strictly speaking, this wasn't allowed) as he'd been away at the time. On the assessment day, after a brief talk from Roger about the dos and don'ts of handling, we headed out to meet the wolves.

Kenai and Kodak had been separated from the rest of the wolves, as Kenai had been deposed as dominant female a few months previously. This pair was housed in the old, small enclosure that was situated in a corner of Roger's garden, and to reach the field we were to walk around we had to walk Duma and Dakota past the two older wolves. There was no love lost between Duma, Dakota, and Kenai, and it was Dakota who had led the coup, although I came to appreciate over the next few years that Kenai was a bossy madam - not always trustworthy and a bit of a bully - and the other females hadn't appreciated her leadership style. And now they were arch enemies.

Watching the senior handlers struggling - and failing - to hold back the wolves as we passed the enclosure, I wondered if my tiny, four-feet-eleven frame was up to the task ... Growling and snarling, wolves and handlers flew by Kenai, with Kenai grabbing the fencing and pulling, causing it to move a scarily long way. The newbies were told to hang back, as wolves, like dogs, can redirect aggression if they cannot reach their target, and I didn't fancy those jaws around any part of me, that's for sure!

Once in the field the wolves quickly calmed down and walked along, sniffing madly, glad to be out and keen to set a fair pace, which meant we struggled to keep up with them. The other assessees looked a little apprehensive but, being used to dogs, I was loving it. When the wolves stopped I boldly asked if I could stroke Dakota: as we had met before I had no fear of her. I approached her and reached down to stroke that soft, warm belly.

It's a fact that the harder you rub, the more they seem to like it, and, if enjoying the contact, will often drop their heads, close their eyes, and spread their hind legs to allow better access (the latter can also be seen when one wolf approaches another to sniff this area). As I caressed Dakota's belly she really relaxed and enjoyed the attention. I slipped in a few Tellington TTouch body work moves as well, just to see if the response was the same as last time. It was.

Later in the walk I was handed the lead of Duma, Dakota's sibling. The wolves were walked on a metal check chain, similar to those used in old-style dog training, and a six foot chain lead, necessary because of their bite power. Duma, for the most part, ignored me at the other end of her lead, which was probably preferable to her doing anything else! Any boring interaction with a wolf is a good one: the alternative can give you a massive adrenaline burst, and get you into serious trouble.

Duma walked quickly uphill, and I soon found myself ahead of the others without a fully qualified handler within arm's reach if anything went wrong. It occurred to me that this wasn't the smartest move but, with no idea how to stop her, and enjoying myself too much in any case, I carried on. At the top of the field we met up with Dakota and her handler, who had gone a different way. I asked if I should let them interact and was told 'yes.' Wolves always go through a big greeting and re-bonding ritual when meeting, and via muzzle-holds, growling and posturing, Duma ascertained her superiority over a crouching Dakota.

As we walked back to the centre, I overheard Roger discussing me with another senior handler, commenting that I seemed very confident. I smiled to myself, hoping this meant that I'd passed the assessment. Later, back in the observation room, the main, on-site visitor building, I learned that, yes, I had been accepted as a trainee handler. This required that I be present at least two half days a month, and more, if possible, in order to build and maintain a working relationship with the wolves, and learn about wolves in general, and handling more specifically. I looked forward to the new challenge, which couldn't have come at a better time for me.

Later, we were taken into the main enclosure to see how we coped with the attentions of the wolves, and, more importantly, how they reacted to us on their territory. It was a cool day and muddy underfoot. Alba, Latea, Lunca, and Duma rushed forward to greet us. Dakota, now the omega, the lowest-ranking wolf, hung back, to avoid attention from Lunca, who had finally taken leadership of the girls after Kenai was ousted. Dakota had briefly come out on top after the battle with Kenai, but Lunca, ever the pushy, in-your-face character, had quickly taken charge, and Dakota had to watch her back to avoid being picked on by Lunca and the others.

As I was the smallest there, the Euros – Alba, Lunca and Latea – came full force at me, testing my resolve and any weakness. I stood my ground as they bounced all around and at me. I quietly but firmly deflected them, and, when I could, stroked their bellies to calm them. The test was over very quickly: I'd won that round, but would the meeting continue in this positive fashion? After the initial exuberant welcome and fuss, most of the pack wandered away, as wolves will do, and I was then able to say 'hi' to Dakota who had been hanging around the edge of the group. We enjoyed a few minutes communing before it was time to leave.

I was invited to the pub for lunch, which we used to do a lot between the morning activities and the members' walk in the afternoon. I went into the toilets to wash my hands, and, glancing in the mirror, discovered that half my face was plastered in mud. Nobody had bothered to tell me; I supposed it must be a regular thing amongst wolfie people. Although it was kind of embarrassing I didn't really care, as I'd loved the entire morning.

A pattern had been set for the next ten years, during which the Trust would come to feel like home; its people my family, and the wolves my soulmates.

Building relationships

Over the following few months I continued to visit the Trust regularly, and began to get a sense of the pecking order, not just amongst the wolves, but the people, too. I guess all organisations have their fair share of strong personalities, and the Trust was no exception. I began to long for Roger to learn my name but, once he did, there were times I prayed he'd forget it again. Don't get me wrong, I loved Roger, with all his faults. Without him the amazing opportunity I'd been given just wouldn't have occurred.

That the Trust was just ten minutes from where I lived was a huge bonus, and also that my job allowed me to become more and more involved, at the weekends initially, but later, in the week, too. Roger saw something in me, and encouraged and coaxed me to get to know and understand the wolves, as did John. Both men were very strong characters, and bloody difficult to work with at times, but they also gave their knowledge freely, and in time taught me the difference between wolves and dogs. I'd like to think I taught them a few things in return, especially on the health and welfare side of things.

I developed a healthy respect for all of the wolves, learning who I could trust and who to be wary of. Lunca and I had an 'interesting' relationship, and it wasn't until years later that I felt she finally accepted me. In my heart, Dakota will always be my favourite, the first wolf I stroked, and Duma, her sister, became a firm friend, too. Alba, although aggressive to and intolerant of a lot of handlers and people, always liked me, maybe because of how I held my nerve that first day in the enclosure. I just don't know but I never, ever felt threatened by him and we had a good relationship, especially after his accident.

Latea trusted me, and I became one of her main handlers on walks when she needed support and encouragement, as she was shy and often a little wary of the public. Kodak and Kenai I had little to do with at first, but, over the years, I nursed both of them to the end of their days, and was the one who sedated Kodak the day he was put down; the first and last time I entered an enclosure with him.

The cubs, Torak, Mosi, Mai, and Mika, will always be my babies, as those handlers who helped rear them also think of them. Collecting the three girls from Dartmoor Wildlife Park will always be a special memory, and rearing the cubs rates in my top ten of lifetime achievements.

I don't remember a huge amount about the first few months after I joined the Trust. My involvement began gradually and slowly, as my old dog, Buzz, was still with me. I do remember he used to come up to the Trust with me sometimes, and would team up with Ollie, another handler's dog. The wolves tolerated handler's dogs quite well, seeming to enjoy the interaction, though we were under no illusion that if they met face-to-face our dogs would be toast. The dogs definitely provided stimulation for the wolves as they fence-ran together, sniffed each other through the fence, or scent-marked where each other had. That there were no safety barriers was a little unnerving for me at first as, being a terrier cross and an independent type, Buzz had a tendency to wander off. Kody or Kenai could so easily have grabbed him through the fence. Life was easier for everyone once the new safety fencing went up.

On my journey to full handler status and beyond, many amazing things happened, and are recounted in the following chapters. Thinking about them stirs some lovely memories, and sadness, too, that those days are gone. My nickname of 'Wolf Woman' is not as apposite, these days ...

UKWCT's aim is to deliver education, promote conservation, and enable research for both wild and captive wolves, with excess funds going to wolf projects all over the world. In my time there I was privileged to be in contact with many world-renowned wolf biologists, and cutting edge research gleaned from them I passed on to the public as the UKWCT's education officer.

To raise funds for wild wolves, we organised and became involved in a wide variety of public events, which ranged from members' walks with the wolves to attendance at county shows; running howl nights, children's events, seminars, and film work with the wolves. As an organisation we had a good reputation for professionalism, and were the only ones permitted to transport wolves into London. Due to our big enclosures and socialised wolves, we were a firm favourite with TV companies requiring pretty shots of wolves in a natural setting. The weekday and weekend event schedules were always busy, and a lot of volunteers were needed to staff them, and this is what we were there for: to increase awareness about the plight of wild wolves, and fund raise for them. The name of the game was education, education, education.

Taking part in hundreds of events, the stories and memories I have as a result are numerous; here are some of my favourites ...

Early morning howl session

I loved film work. A lot of footage exists of our wolves looking off to one side whilst a TV presenter interacts with them: what they are looking at is me, just out of shot, usually with a hotdog in my hand. One of my favourite memories is of an early morning shoot for the TV documentary, *Pedigree Pets Exposed*. The light was fantastic, just post-dawn, and Torak, Mai and Mosi were behaving beautifully. Film directors were often astonished by the things we could get the wolves to do, asking, a little self-consciously, if the wolves could do this or that, to which I'd reply "For a hotdog, they will do pretty much anything." It's easy to get wolves to howl, especially those we had reared who, when cubs, looked to us for advice in the art of vocal communication. The required shot was the pack on top of the mound, howling.

I ran up to the top of the mound, called the wolves to me, got them howling, then dropped out of shot by laying on the mound behind them. The result can be seen in the programme. Should you see it, think of me lying in the wet grass, about to be pounced on by three excitable wolves!

I was often asked to be in front of the camera, too, which, I must confess, I adored. I think my first experience was for the hit TV series *Talking to Animals*, which featured Sarah Fisher, the UK's top Tellington TTouch instructor, working with a variety of animals using the Tellington TTouch Training method. As an experienced Tellington TTouch practitioner myself, Sarah asked if they could film the wolves with me for the series. After discussions with Roger, John, and Rich Watts, another senior handler, this was agreed, and we took Duma and Dakota into the photo enclosure, where Sarah and I talked about and worked with them.

Sarah and I were fine, but, of course, the nippy nickers, Duma and Dakota – so-named because of their propensity for grabbing anything novel, and delivering sharp nips if prevented from doing so – were all over the camera and boom. I was in charge of protecting Sarah, who was a natural with the wolves and had no problems, whilst John and Rich defended the cameraman and his equipment. Needless to say, by the end, both John and Rich had minor collateral damage, mostly to their thumbs. I remember John diverting Dakota to one side, away from the camera, and as she sailed past Rich, she grabbed his thumb. All was taken in good part, although I got several weeks' ribbing about sore thumbs and being a diva.

We spent one summer filming an in-house video of the wolves to promote the UKWCT and our work. The result was hours of footage, a whole host of outtakes, and a stunning film. With calmness and patience, I was able to achieve what the director or cameraman wanted with the wolves. Time, patience – and a lot of hotdogs – usually did the trick, plus my really good relationship with the wolves. It was generally the most experienced handlers who did the film work as we knew how close we could let the wolves get to the crew and presenter, and how to control and maintain calm with both them and the people we were working with.

And it wasn't just film work but radio, too. Getting a wolf to howl on cue on live radio was always a little hairy, but fun ... so much fun. Out of all the events we ran, I loved film and radio work the best.

The power of the howl

As education officer, I ran school visits, where we either took the wolves to the school or the children came to the centre. The kids simply loved these events, and we had every category of child: troubled, autistic, disabled, and mainstream.

The troubled kids would arrive, hoods up, non-communicative, not even looking me in the eye. I'd chat to them, about how a wolf pack is structured exactly the same as their families were, and talk to them about how wolves deal with conflict. The truth is wolves are very peaceable creatures: within their family units none of them wants to injure another, and this makes them good conflict-resolvers. I hoped that the wolves' example would let the kids see that they had a choice when it came to how they reacted to volatile or unhappy situations.

By the end of the visit, usually, the hoods had come down, and the kids were smiling and interacting. Many did wonderful art work or creative writing afterward, which helped in their therapy. I often wonder about those youngsters, and hope we helped turn their lives around in some small way.

On one particular school visit at the centre, after I gave the talk about wolves in general, we walked around the enclosures as I chatted to the children about our animals. Towards the end of every visit, I always got the kids to howl in the hope that the wolves would respond in kind. This particular time, on the way back to the education room, I remember one small boy chatting away to me, and we talked all the way there. After their coach arrived to take them home, a couple of the teachers who had accompanied the children told me they had taken photos of me and the little lad conversing. When I asked why, they told me that he was autistic, and didn't talk to adults; they'd taken the photos to show his mum. We all had tears in our eyes that day. The wolves' ability to bring out something in kids was incredible: they could reassure a scared or ill child, help troubled kids see a brighter future, and so on. It never ceased to amaze me that they did this; they were true ambassadors.

The Rosebud incident

Sadly, Roger Palmer died in 2004, and this was a turbulent time for all of us, with so much change and uncertainty. The funeral had been a family affair, but Roger was an influential person, and many people wanted to attend the memorial – which Roger had requested the wolves be present at – held in the local church.

The church was attached to Bradfield College, a huge private school, and next to the church was a large lawn. Roger was a keen hunter, and his drag hound pack and Beagle pack were also scheduled to attend, and all of the UKWCT volunteers wondered what the day would bring.

As Duma and Dakota were the easiest to handle, we decied that these two should represent the pack. The trailer to transport them was parked as far away from the dogs as possible, and we quietly collared the wolves and brought them out after the service. We had lots of extra help, bearing in mind that a great many of us wanted to attend the service. All of the dogs were off-leash, and the drag hounds seemed totally disinterested as they ran around on the other side of the lawn. Occasionally, a Beagle would break rank and begin running towards us, whereupon one of us would intercept the animal, pick her up in mid-stride, turn her around, and set her down to run off back towards the pack, the dog barely registering she was now running in the opposite direction! It was hilarious, if a little unnerving. I distinctly remember the kennel master shouting in a very loud, low, gruff voice "Rosebud, Rosebud, come here," which seemed such a feminine name to come out of such a big, burly man. Rosebud, you will be glad to know, lived to hunt another day.

Fur and leather

As mentioned, the centre held many public events, and sometimes we were booked to give talks, and meet and greet at nature reserves. Two incidents at events such as these stick in my mind; both times I was handling Dakota.

One busy day, Dakota was taking a break from meeting her adoring public, although Duma – never tiring of attention from the crowd – was still at it. Dakota lay down in the middle of all the mayhem and took a nap: not only did she have her eyes closed, I could see rapid eye movement, too. Don't be fooled, though: even in this state wolves are super-alert to everything around them. One second Dakota was sparked out in REM sleep, the next she was up, her snapping jaws just millimetres from the fur trim of a girl's coat as she walked past, a little too close to our sleeping wolf. I just managed to stop her locking on to the coat. As handlers we were taught never to take our eyes off the wolf; this near-miss proved why that instruction was so important.

On another day it was a man who caused my heart to skip a beat, and not for the right reasons. When people meet and greet a UKWCT wolf, they approach from the front so that the wolf can see them coming, and we can judge from feet away if they are happy to interact with the approaching individual. Once within range, the individual is asked to extend their fist for the wolf to smell, which might be a cursory sniff or more if they find the smell interesting (many's the time a wolf has unceremoniously shoved his or her nose in someone's crotch).

Today, though, Dakota was very interested in a gentlemen's pocket, and I asked if he had any food on him. "No," he replied, and promptly extracted his leather wallet from his pocket to show it to Dakota, shoving it right in her face. My instantly reacting to pull her back saved the man from having to explain to his bank why new cards were needed, and let me tell you, credit card companies refuse to accept this explanation, as a UKWCT handler who had to have that conversation discovered! The poor man appeared shocked at my sudden reaction, though when I explained that leather is edible to a wolf – and rather tempting – his expression changed to sheepishness.

HOLDING HANDS

Lunca and I had a rather rocky relationship in the beginning, as noted earlier in the book. I think she realised I was wary of her, which exacerbated the problem. When I first became involved with helping John with the wolves' welfare, the Euros had an ongoing health problem that nothing seemed to cure: recurrent sore feet, which no veterinary drug, wash or ointment seemed to alleviate.

I knew many alternative health practitioners, and, luckily, John was into complementary therapy. When I sent the wolves' hair to be analysed, it turned out they were suffering from a fungal infection brought on by spores in the spring grass, the cure for which was to treat them systemically with nettle root capsules and ionic sulphur, which we'd mix into meatballs and feed them one each a day: by hand so that we knew each wolf had received the correct dose. After a while, Lunca seemed to develop an obsession with my right hand, which I normally used to give her the meatball, and began holding my fist on public walks, taking me with her as she marched around. She never really applied pressure but my hand did become very cold and sore.

After a walk one day, with the public watching we entered the enclosure, and Lunca almost immediately grabbed my fist ... which was fine until she transferred it to the side of her mouth and began to bite down. To counteract this, I did some

Tellington TTouch body work on her muzzle and inside her mouth, and the pressure eased off. This ritual went on for a while until Lunca eventually lost interest. Exiting the enclosure I checked my hand, which was painful, the skin tight and shiny, with bruising beginning to appear. I went home and took loads of arnica and did some Tellington TTouch work on it, to ease the pain and bruising.

When she did this, the trick with Lunca was to remain calm and not react, and certainly never show fear, or tell her off. She was such a forceful character; if you didn't have a really strong bond with her, it was not possible to reprimand her. Happily, she did mellow over time, but, as a youngster, boy, was she challenging to work with!

If you go down to the woods today ...

Walking the wolves along permitted paths on part of the Benyon estate was a highlight of the week for wolves, handlers, and members of the public alike. They were not without their dramas, though, like the day a dog got through our lines and ran straight towards the Euros. Luckily, the dog chose Latea to sniff noses with, and she just wagged her tail, the dog running off again before Lunca and Alba arrived.

The woods were not fenced off, and anyone could walk the paths, which made for interesting encounters on occasion, even though we'd have people walking point to spot dog walkers, or any other potential problems. They sometimes had to politely ask dog owners to put their dogs on a leash for a while; if they hadn't seen the logo on our jackets, they might ask why, or even got a little antsy. Generally, however, when told that a pack of wolves was approaching, they quickly complied.

My favourite encounter in this respect was with a guy who regularly ran in the woods with his Jack Russell, who didn't even break stride when he reached down, scooped up his dog, ran past, and returned his dog to the ground in one seamless motion. The dog never even looked back, either, just continued on. It was great to watch.

Dogs were not the only other species we encountered, either, and Alba's favourite were horses. Moving off the track to let them past, Alba had a habit of standing up on the shoulders of handlers to watch them go out of view, a "Is that dinner?" look on his face. One day we even encountered the local hunt passing through, dogs and all. The Euros loved it, getting so excited that they howled, which they never usually did there as they didn't regard the woods as part of their territory. I think it was the hunting horn that set them off.

One day, whilst I was handling Dakota, four or five deer came running towards us up a wide firebreak in the woods. This was in the days before we double-leaded the wolves (two leads attached to the animal, with two handlers holding them at all times in case one of us fell over or dropped the lead), and I shouted for back-up in case Dakota took off after the deer. She was absorbed in sniffing the ground, and, as the deer came closer, they diverted off the track and into the trees. It was only as they disappeared that Dakota looked up and excitedly sniffed the air, her expression clearly saying "I think there have been some deer nearby recently." I didn't know whether to laugh or cry for her, at the same time relieved that I didn't get dragged through the trees backwards, hanging onto a powerful wolf in hot pursuit of dinner.

The wolves often ignored or missed game out on walks, and my feeling is that they were too well fed to bother, or simply knew they wouldn't get very far in a chase with us attached to them. The Euros used to try and grab pheasants, though, on walks, and we would 'lock on' – stop dead in our tracks in an effort to prevent them moving forward – and then just wait until they lost interest. The Euros had a habit of appearing to comply by walking away, but then would whip back around behind you in full flight again. They really did keep you on your toes.

All of the wolves loved water, and near the end of the two-hour walk, we would stop at the lake and let them take a dip. Footing, here, was difficult for handlers, and the wolves often got really excited, especially Alba and the girls. We would often attach another six foot chain to their usual chain to give them twelve feet of lead, which would allow them to get further out and have a swim.

The Euros used to team up, find an object – such as a log at the bottom of the lake – and roll it back to the bank. After he got out, Alba would excitedly charge about, his handler barely keeping up with him, whilst the girls liked to sidle up to some poor, unsuspecting handler and have a good shake!

Alba especially loved water, and on hot days would lay down in all the puddles he could find to cool off. During hot weather – when even the big puddles dried up – he would still lay down where they should have been. Habit or did he not realise? I don't know, but it made me laugh to see him do it. Roger's favourite saying was "Wolves are creatures of habit" and he said it at least four times a day.

SHOWTIME

Every summer we had a busy schedule of taking the wolves to county shows and other big outdoor events, for which we'd have to erect a huge travelling cage, which took us at least two hours to do. This enclosure incorporated a double-gated, airlock system to enable us to get in and out without the wolves escaping, plus an overhang so they couldn't jump out. Often the team erecting the cage would leave early, and John or Rich and I would follow along after with the wolves.

It always made me chuckle to see folk do a double take when they read the 'Wolves in Transit' sign on the trailer. Moving wolves around necessitates a lot of paperwork, and occasionally we would be stopped by the police and asked for this or that form. Usually, the police admitted they had only stopped us because they were interested and wanted to see the wolves, and we always opened up the side door so they could meet them. I wasn't there at the time, but, apparently, the trailer was once stopped on the Royal Mall in London, on the way to do a TV show. This drew quite a crowd, and the handlers were handing out leaflets about UKWCT.

The situation wasn't helped by Torak's tendency to stand up on his hind legs and look out the back grille every time we stopped at traffic lights. I bet the faces of the people in the vehicle behind were a picture, if only we could have seen them! If I was travelling in the back-up car directly behind, I always used to wave at him and he seemed to grin back at me.

Once the travelling cage was up at a show, the wolves would generally sleep the day away whilst being admired by the public, or else might take part in meet and greet sessions. Arena events were fun. I remember walking through Newbury Showground with Lunca, Alba and Latea, and having to ask the fairground firing range to hold off while we went past, so it didn't freak out the three of them. Once in the arena we would walk round, whilst one of us would talk over the public address system. Sometimes, for a little variety, we'd spray perfume on the ground so that the wolves would roll in the scent.

One time, when Torak, Mai and Mosi were about a year old, we took them to Highclere Game Fair. It was a horrid, rainy, cold May day; the poor darlings, having begun their winter coat moult, were freezing. We relented and put them in the trailer in the end, but not before an arena event that had been going really well until we were leaving the ring, when the speaker screeched just above Mai's head. She really didn't like things like that, and stopped dead. There was no way she was going to walk past the speaker, and in the end, in front of hundreds of people, I had to pick her up and carry her all the way back to the stand. I just smiled and gritted my teeth: even though she was only a year old, she was still pretty heavy. It's a good job we'd made a point of picking them up as they grew, otherwise she may not have let me carry her. It nearly killed me!

It wasn't all work, work and more work at the shows, of course. Burger vans often donated old burgers and sausages for the wolves that they couldn't sell, and they loved an ice cream as a special treat. Handlers got some time off, too, to wander around. If it was a two-day event, we sometimes even camped out.

Research

Many students passed through our doors over the years. Some came and went with little impact, whilst others proved their worth and remained in contact. Part of my job as education officer included guiding and advising students in their dissertations. Some had unrealistic ideas of what was achievable, whilst others took on board my advice, or became interested in subjects I wanted answers to.

Duma, for example, had a passion for eating thistle flowers, delicately reaching up and nibbling the heads, especially when they went to seed. I could never figure out why she did this. Wolves are very good at self-medicating, but whether or not she gained anything from this I don't know. It was fascinating to watch, though.

We expanded on this when Vicky Allison, one of the students, and later my assistant education officer, did a paper on self-medication. Vicky designed a wolf-proof container, akin to a heavy, metal water trough with a fixed mesh top, in which were planted herbs – rosemary, basil, fennel and sage, amongst others. As they grew, parts of the plants poked through the mesh, and the wolves would nibble off the tops without stealing, destroying, or rolling on the whole plant (wolves love novel smells, and the plants would have been trashed in seconds if not protected).

As expected, all of the wolves went mad for the herbs to begin with, and, over the coming weeks and months, preferences became evident. The older wolves especially, I remember, loved the rosemary.

Over the years we conducted research into many things, including food intake in different seasons. Wolves eat relatively little during the summer, staying cool by remaining lean and having only one coat layer (the outer guard hair), after losing all of their soft, warm, insulating under-layer around May time: the only time a handler gets covered in hair.

My favourite memory, though, is of trying to collect saliva samples from the wolves. I had a phone call one day from a university that wanted to compare DNA in different wolf species, if I remember correctly:

"Could you take a swab from inside the wolves' mouths?" I was asked.

"I'll try," was my response.

All of the wolves would regularly come up to the fence, for a fuss or occasionally a special treat, so something new in my hands – such as a swab stick – would always pique their curiosity. Armed with a few spare sticks, I went on to the yard of each enclosure and called the wolves. The trick was to get the swab stick in a cheek without the animal grabbing it and pulling it out of my hand, and I couldn't believe my luck with the first pack, it was so easy: they sniffed, they licked; I was able to wipe around inside their mouths a bit, then retract and bag the sample without any problems at all. Easy!

Then came the turn of Duma and Dakota.

Duma wasn't too bad: she did try to grab the stick a couple of times but I was able to hold on to it. Dakota, on the other hand, more than lived up to her reputation of being difficult ... A wolf's jaw is so strong, the minute they lock on and pull back, it's like being attached to a ten ton lorry that's reversing. The sample stick was whipped out of my hand as if it had butter smeared all over it.

I tried again.

This time, she only nibbled on the tip of the stick with her incisors, the very front teeth. I think I lost two more sticks, but eventually she became bored with the game and I was able to get the sample.

Once all of the wolves had wandered off I was able to retrieve the dropped sticks, chewed and broken, but not, thankfully, inside Dakota. I hope the university scientists were grateful – I never did learn the results of the tests!

I have so many memories of public events; photo days when the Euros used to become bored and play 'Let's bounce the handlers' (jumping all over one of us until we had to be rescued by the others!); Christmas Cracker events where we would dress Christmas trees with edible crackers; howl nights where the wolves stole my thunder by howling so much that I couldn't give my talk.

And wolves looking in through the education room window at the people inside: "Look at all the funny humans in a box," you could almost hear them say.

Ten years and ten thousand memories – and they're just of the wolves, never mind the people I met along the way.

My fondest memories, though, are of after hours, when the show packed up and the public went home. That's when the real fun began ...

Behind closed doors

Although I loved the events we held, and educating the public about our wolves, it was those special moments when only a few of us were around after the public had gone home that I cherish the most.

In that surreal interval of crepuscular light, magic happened as the wolves came alive: playing, hunting, singing. At their most spirited, it wasn't safe, sometimes, to enter the enclosures, though okay to do so when they were calmer. To stand in a pitch black enclosure, trying to pinpoint a wolf coming in for a feed, only to have him run past you and disappear just as silently as he had approached, was breathtaking and dream-like. My strongest memories are all of this time of day; these are some of my favourites.

Night-time

When I first started at the UKWCT, initially helping out from time to time only, tagging along for fun, all of the wolves were housed in kennels overnight, for security reasons and to keep the noise down because of the neighbours. Eventually, I regularly put the wolves to bed, and my first memorable encounter occurred in August 2002.

First feed

Just over a year after I started as a volunteer at the UKWCT I found myself, after a long day of events and a night in the local pub, agreeing to help put the wolves to bed. John, Rich, another senior handler, and John's wife, Paula, and I drove up to the centre and got out of our cars in darkness. We had very little security lighting then, and torches were needed to find our way to the enclosure and kennel light switches.

Arriving at the Euros' enclosure we found them all at the gate, waiting for us. I was still a bit wary of Lunca, and often opted not to enter an enclosure with her, but because at dinnertime you can pretty much guarantee their minds will be focused on food, I figured it wouldn't hurt our relationship if I helped feed her.

Once on the yard, John opened the gate and all three – Alba, Lunca, and Latea – piled in. I stood on the yard with my back against the fence: you don't want to get knocked over, or inadvertently take a step back around wolves, so a sturdy fence behind you is always preferable.

Lunca and Latea ignored me in preference to the others, but Alba came straight over, jumped up at me, and took the whole of my head in his mouth. This is standard procedure in wolfie greeting etiquette, I now know, but, unused to it then, in-between holds I quickly checked his body posture (and especially his tail and ears) to reassure myself it was a true welcome and not a prelude to something more sinister – Alba had a reputation to uphold, after all, and I was still a newbie. All was good, however, and I received not just one muzzle-hold but three; each time, practically my whole head disappearing into his red maw. His meaty breath was something else, though, especially for a vegetarian. At one point our teeth even clashed, though I'm pretty sure that Alba didn't feel a thing (I did). With teeth still rattling, Rich told me I'd done the right thing by letting Alba greet me.

Alba remained on the yard – typical boy, always ready for dinner – but Lunca and Latea, not ready for bed just yet, dashed back out for another circuit of the enclosure. John, Rich, and I gathered up leads and collars and made our way into the dark in hot pursuit: well, stumbling and tripping is a more accurate description, our night vision nowhere near as good as that of the canids.

John caught up with Lunca pretty quickly, and with the aid of a reassuring belly rub, got the collar and lead on her. With his usual sense of humour, he rumbled "Well, I've got my wolf." Rich disappeared into the wooded area just as Latea came out the other end. A couple of rounds of the enclosure later, he and I were no closer to catching her. As she normally followed us to the yard, we headed that way, leaving Lunca and Alba contained on the hard standing whilst we put the other wolves to bed.

We were quickly able to get in Kody, Kenai, Duma, and Dakota, and, sure enough, Latea was waiting by the gate as we walked back. Once all three were on the yard, John asked me to open the door to the 'restaurant.' Alba, having had to wait the longest, had become over-excited by this point, and raked – and my hand – as I attempted to prise it open, with all three fully-grown wolves trying to get past me. The door opened outwards so it was always a tussle. In his impatience, Alba nipped my elbow as if to say "Hurry up!" and, lo and behold, I finally got the door open. In a flash they were in, and Alba finally got his dinner.

That night established a routine. John was always up for socialising, and, after a hard day, various combinations of us would retire to the pub, then put the wolves to bed, often taking leftover food with us as treats.

Full moon

People often ask how wolves behave around a full moon, due, no doubt, to them often being portrayed as beasts who howl at the moon ... but this isn't the case.

Wolves howl as part of a complex communication system: a way of keeping track of each other if separated, defending territory, finding a mate, or simply for social bonding. Wolves love a good rally chorus just before they set off on a hunt, and, as they don't hunt in the dead of night, the only reason they would have to howl then is if something sets them off. For our wolves that would be anything two-tone – emergency sirens, ice cream vans, and even the local church bells!

One hot summer's night, under a beautiful full moon, I found myself at the centre with the usual crew after closing time. Wolves are really good at self-regulating their body weight, and will eat less in the summer months as they don't need to put on or maintain a fat layer to keep them warm. We would starve them one day a week anyway, and when coming up to a food-free day, it was sometimes a struggle to get them to go to bed.

Under bright moonlight this particular evening, the Euros especially didn't respond to our arrival, choosing to remain laying out in a shallow scrape (a hole they dig to access cool earth to lay in), in the grass. I remember they were particularly calm, so we entered the enclosure and made our way over to them.

None of the wolves even got up, just rolled over, indicating they wanted a belly rub. We first sat and then lay down next to them, quietly chatting and giving the wolves attention if they asked for it. Mostly, though, we soaked up the atmosphere and enjoyed the moment. On the whole, it's inadvisable to sit or lay down around wolves, but, with the right people around you and the wolves so chilled – and if you have the right relationship with them – it's the best feeling in the world.

Mostly, though, feeds were a manic affair, with wolves scrambling to get their share of the food left for them inside the kennels. We would let them competition-feed (as they would in the wild), in the knowledge that, if a particular wolf got more than his or her fair share one night, they would be less competitive the next, allowing others to take the lion's share. Generally, as I say, the wolves were eager to enter the kennel or yard, which is not to say that it always went smoothly ...

One wolf in, two out; black wolves

If any pack was to cause problems, it would be either the Euros or the Brat Pack (my nickname for the cubs, Torak, Mosi, and Mai, who I helped raise). Both packs were renowned for not being in unison: two of their number might be on the yard and one outside, but just as the latter entered the yard, one of the other two would rush out. As the kennel yards had several gated hard standings sandwiched between the enclosure and the kennels, occasionally, it was possible to slip into the next yard, open another gate whilst some of the wolves were contained safety next door, and get in the elusive troublemaker. Once all were reunited on one yard, it was a matter of opening the kennel door and letting them rush in together.

Two things could go wrong with this well-thought-out plan, however.

Firstly, the Euros could open gates. All of the gates could be bolted on each side, so if you forgot to secure the lock on the outside, the wolves were able to let themselves out again. John first discovered this when they were young cubs, when one day he found himself taking one cub back to the temporary artificial den from an enclosure, only to discover the others happily trotting along behind him! Everything was double-locked and padlocked around the Euros, and with every pack as a precaution – especially the gate leading off the yard and into the outside world.

The second possible hiccup occurred if food was too close to the door the wolves were let in by. Wolves would not naturally feed and sleep in a kennel, obviously, and would eat alfresco by choice. If one of them was able to grab a chunk of meat from within the kennel, and reverse out of the door before the others went in and you could shut the door behind them, you had a problem. It isn't like you could tell the offender to drop the meat (which, by this point, was probably already in their belly, as wolves can swallow huge pieces of meat without chewing it). You also couldn't then simply reopen the door and let them back in as the others would rush out. It was a real dilemma.

Torak was an expert at this trick. He could hover half in, half out of the enclosure gate, seeming to want to go in, but then reversing at speed. This meant that there was then too much food in the kennel for just Mosi and Mai, creating a problem the next day, as the girls wouldn't be so hungry (having eaten Torak's portion as well as their own), so reluctant to come in at feeding time. I took to not putting the full amount of food in the kennel, and posting more under one of the trapdoors once Torak was safely secured on the right side of the kennel door. This was a little risky for me and my foot, because, as I kicked it under, three pairs of snapping jaws would be trying to grab the food from the other side!

And this wasn't the only trouble that the Brat Pack caused me.

Before they grew used to the routine, as youngsters, they did their utmost to avoid going inside to feed, wanting just five more minutes of play. I remember one night, when they were around five months old, stumbling around a dark enclosure, with Rich and Alex, another handler, hunting for them for what seemed like hours, not wanting to leave vulnerable - and certainly mischievous - cubs alone and probably up to no good in two-and-a-half acres. Alex, even today, takes great delight in reminding me of my frustrated outburst that dark night: "Whose idea was it to get black wolves?" Being outwitted by wolves was a regular occurrence: they are super smart, very strong, and very fast.

In later years, after Roger had died, and John's health began giving him problems, I took on the task of feeding the wolves one week in three or four, depending on who else could do it. We wanted the consistency of one person feeding for a complete week at a time, in order to monitor the wolves and judge which day to starve them, and so on. We all had our own routines and ways of doing things as well, which the wolves came to know and expect. Wolves hate change; it unsettles them, and although it was difficult sometimes to drag myself out of a warm house late at night on cold winter evenings, once I arrived at the centre and saw their expectant faces at the fence, as they ran alongside my car, it always felt worthwhile. I loved it.

Admittedly, it *was* a little scary being up there in the dark on my own, but I figured if anyone else was on-site, the wolves would tell me. And if the worst came to the worst, I'd just step into the enclosure with them. No sane person would follow me in, surely?

Fireworks and barking deer

Wolves can be inquisitive by nature, and novel things either scare them half to death or pique their curiosity. Two things were guaranteed to distract them: Muntjac deer barking - an eerie noise, suggestive of a dog being strangled whist barking - and fireworks. Both noises fascinated the wolves, and they would come close to the yard gate, then run off again for another listen or look. They weren't frightened of the fireworks - they didn't bark, tuck their tails or crowd near me for reassurance - and seemed simply to like watching them, the sight and sound sending them off on another circuit of the enclosure. With the Muntjac, they'd listen intently and sniff the air, picking up scents on the breeze. On nights like this I'd simply leave them to it, even though it mucked up my 'score:' how many nights each of us managed to get every wolf to bed brought out the competitive streak in our natures; the ribbing you got if you failed was merciless.

Whatever the reason for being at the centre at night, I always loved it. To really know wolves it's necessary to observe them in every phase of life, time of day, and each passing season. The changes I saw from even one hour to the next could blow me away: sleeping one moment; haring around, playing the next. Wolves love to frolic, and dawn and dusk are the times to see them at their best.

Dawn

Although night-time encounters were great, mornings were my favourite times of day with the wolves, when it was rare for anyone to tag along with me so early (6.30) in the morning, like they sometimes did at night. This special time I enjoyed with the wolves - the beauty of the sun rising, and a new day beginning - always calmed me. Yes, there were times when it was still pitch black, but, as the light gradually seeped into the night sky and a new day dawned, it brought with it an amplified awareness of the changing seasons, and helped me to feel more connected to nature.

I became involved with letting out the wolves when John had a bad foot, and I offered to help temporarily. Once he had recovered, however, I persuaded him to let me continue, and we shared the task, although he would usually turn up at some point to walk his dogs, and check on me and the wolves. We'd have some of our best conservations over an early morning cuppa, or sneak a cuddle with the wolves when it was just the two of us, with no one else around. We'd also walk John's dogs down past the enclosures and around the bottom field that belonged to the UKWCT.

Kenai and Darcy

One of John's dogs, a rescue Collie called Darcy (so named by the rescue centre because he was stunning to look at, but ever-so-slightly dangerous), had a game he played with Kenai. Once the wolves were let out and Darcy was released from the car, he'd run at top speed alongside Kenai's enclosure with just the safety fence between them. Kenai would keep up but appear to be ambling along at half speed. Towards the end of the enclosure, or when he decided, Darcy would bark as a signal, and both would turn and hare back in the opposite direction, and this they repeated time and time again, until Darcy was too tired to continue, at which point Kenai would saunter off with a look that said "Can't hack it, eh, pup?" She was never winded like Darcy, or looked like she was in predatory mode; just seemed to enjoy the game. Of course, we were under no illusion that, if we'd put the two together, Kenai wouldn't attack Darcy, because the chances are she would. The security of the fence meant they were both able to enjoy their early morning workout without any cause for concern.

Kodiak (Kody)

My fondest memories of Kody originate from early mornings.

Towards the end of my time at the UKWCT, I persuaded the directors to allow the wolves to stay out at night, with access to the kennels via the trap doors in case they wanted to shelter from bad weather. Consequently, we changed their feed time to early morning, which meant I'd often still be at the centre at around 7am. On sunny spring and summer mornings, at well past dawn, most of the wolves would be up and about playing or hunting for mice. Kody, however, liked to lie in.

I'd take a wander around the outside of the perimeter fence to check out his favourite sleeping spots, often with the feed bucket in my hand, and find him curled up in a scrape surrounded by long grass, dead to the world. Wolves are generally super-alert, and there's nothing that escapes their notice, usually, but I could have dropped a bomb next to Kody and he wouldn't have stirred. I'd call him, clang the feed bucket, call some more, until – finally – the old boy would raise his head, as if to say "Eh? What? Is it breakfast time?" A big stretch would follow, and, depending on whether or not he was hungry, he would run or saunter to the hard standing for his food. Being old, I always liked to check first thing that Kody was okay, and get his meds down him via a tasty sausage or meatball.

Kody behaved the same even when kennelled. With the other packs, I'd generally go onto the yard and let them out via the kennel door, but as Kody was very dominant, it wasn't safe to do this with him, so instead I would open the trap doors from outside the runs. The others would shoot out of the traps, eager to be outside, but Kody would be nowhere to be seen. If he didn't get up, I'd peek through the cracks in the door to check he was alright, and call him to see if he would wake and come out for his meds.

On one occasion I had a very close near-miss. I was always super-cautious about security, and trained myself to always shut gates behind me; check and double-check everything before I went in with or let wolves out. That day my vigilance saved me from a very nasty situation.

All of the four kennels had interconnecting doors, as well as doors that gave entry to the outside and run area, which meant that the wolves could be given more or less space, depending on pack size. On this occasion the wolves were given the middle two kennels, with the outer two used to store straw, leads, and other wolfie paraphernalia.

I went on to the first yard and (to this day I don't know why I did it apart from discipline and routine) put my foot against the door of the first kennel before easing it open slightly to peer in. The minute I did, a loud, menacing Kodiak-growl rumbled at me from the other side of the door, which I quickly slammed shut and locked.

Going back out and looking through a crack in the outer door, I could see Kody strutting around inside, growling at the door that only moments ago I'd opened. The previous night's feeding team had obviously put him in the wrong kennel, and if I'd not checked, I'd have walked straight in with Kody behind me, between me and the door, and no one on site to come to my rescue.

As this realisation hit me my legs turned to jelly and I had to sit down for a minute. I have no idea how Kody might have reacted: he knew me well, and I would hope that my training would have kicked in and saved me, but I'm glad I didn't have to put this to the test. It was a lesson to all of us to check, check, and check again. Needless to say I received an apologetic phone call later that morning from those responsible for the error.

TORAK

As Torak grew from a goofy cub to a lanky teenager, he was much bigger and stronger than me from about six months of age. By the time he was seven or eight months old, I stopped going into the enclosure with the cubs on my own. Wolves, like children, will eventually test to see how far they can go: boundary pushing for youngsters of any species is normal and healthy ... except when you're a wolf handler. Then, it can get you into trouble, big time!

Early one morning, when I went to let out the Brat Pack, I discovered that only Torak and Mosi had gone to bed. Mai was waiting out in the enclosure and, after giving her a fuss, I released the other two. It was only then I remembered I wanted to check Mai's vulva. Wolves are claimed not to be sexually mature until 22 months of age, but I'd noticed that both girls had a greenish discharge from their vulva, and I was keeping an eye on this.

I wandered after the youngsters as they ran and played, followed by some re-bonding ritualised aggression. Torak spotted me and decided it was time I joined in the game, at which point I realised I was a very long way from the gate out of the enclosure.

Torak ran towards me and bounced up, leaping at my head. Being around six foot tall from the tip of his nose to the tip of his tail, he towered over me. I stood my ground somehow and braced myself; he came again and a third time, until finally responding to my command to stop. I calmed him with a belly rub (all of our wolves were used to having people stroke them on the abdomen, which calmed and reassured them. Roger always claimed this harked back to when they were neo-natal cubs, and their mother licked them in this area to encourage them to toilet).

Once Torak was calmer I began to casually wander towards the gate, but he came at me again, and this time I had to pin him down on the ground and hold him there until I could see he'd submitted. It was then that John turned up on-site, and the arrival of the dogs distracted everyone enough to allow me to leave the enclosure. By that afternoon, every muscle in my body ached from the force of Torak's bouncing; even though it was only in play, I had to really brace against the power of it.

The cubs had become increasingly boisterous towards us of late, and I often had them hanging off my clothes as I let them out. Sometimes, they'd howl, and in the ensuing mayhem grab various bits of me and pull in different directions, which, of course, was fine when they were tiny, but not safe as they grew bigger. Until they learnt some manners, I regrettably decided I would have to give up my early morning solo adventures with the Brat Pack.

Torak and the pumpkin

I have only two visible scars from my time with the wolves, both minor collateral damage. I was lucky: I never experienced a serious incident with any of them, or felt the power of their jaws when they really bit down.

One of my scars was Torak's doing. This looks like a freckle, and is sited in the webbing between the little finger and ring finger of my left hand.

As part of my role as education officer, I ran children's events. At the Halloween party (which was always popular) as part of the event, the kids would carve out pumpkins, stuff them with some of the wolves' favourite treats, and I would give them to the wolves, who would rip them open, eat the treats, and even the pumpkin. As a species, wolves are opportunistic feeders, and it is usual for them to scavenge for apples, nuts, and berries to supplement their predominantly carnivorous diet.

The morning after the first event John and I met up as usual to let out the wolves, have some wolf-time, and generally put the world to rights. I wanted the children who came to each event to feel special, so a gathering up of all the old pumpkins was necessary before that day's party. John and I began to collect the leftovers and put them on the yard. Torak, however, thought this was a great game, and stole a piece from me with ease. John, in his wisdom, claimed this didn't make me look good in Torak's eyes, as he might think he'd got the better of me. I'm sure this wasn't the case, but John being John, I had to go back and retrieve said pumpkin. Torak, of course, thought this was brilliant entertainment, and tried to snatch it from me again.

Just as I got to the gate and was about to throw it on to the yard, Torak lunged for the pumpkin piece, accidentally catching the webbing between my fingers, leaving a small puncture wound, probably with one of his incisors. For some reason the scar has never gone, and acts as a permanent reminder of Torak and the pumpkin incident.

Alba and the racetrack

In the summer months we'd let the grass grow in the wolves' enclosure, as this provided an additional hiding place for them, and also small animals such as mice and voles, which, in turn, benefited our resident birds of prey, the kestrels and owls.

Eventually, the wolves tired of the grass, which was harder to run through, and, as it had also begun to look a little scruffy for the public, one morning, John took the tractor into the enclosure to cut a few paths through it.

Once outside, Alba became really excited. He ran to the track John had made, cocked his leg and peed, then took off like greased lightning, not stopping until he had got back to his own scent mark, which he sniffed before running back the other way. Wolves can easily run at 30mph, and some individuals can reach a top speed of more like 45mph in short bursts. Alba was stretching his legs on the newly-formed 'racetrack,' and, unimpeded by long grass, was really enjoying his early morning workout.

I loved the fact that the wolves at the UKWCT had big enclosures: wolves are born to run, and in the space available ours could get to full speed. One of my favourite memories is of Alba revelling in the racetrack.

Dancing wolves

Not only could Alba run, but he could also dance, and this happened in breeding season when he stuck like glue to Lunca's side.

Breeding pairs become increasingly attentive toward each other in the build-up to actual mating, and near the time that the female is about to ovulate, about ten days into her season, the breeding male will be constantly by her side, waiting for the optimum time to mate. As wolves have only one season a year – around January/February – it's essential that the opportunity is not missed.

Alba mirrored everything that Lunca did. They parallel-walked or ran together, nuzzled each other, and Alba would groom Lunca. I remember one cold morning, watching, mesmerised by their display. It was as if they were joined by an invisible cord. They were so tender toward each other, and really in tune as a result: it looked as if their guard hairs were in constant contact. Both animals were completely oblivious to me standing there, and Latea moving around them at a distance, as Alba gazed affectionately at Lunca. It was probably the most private window I ever got to look through in all my time with the wolves. Beautiful, serene, and intimate is the only way I can describe the moment.

Kenai

In the last few years of her life, Kenai suffered from a cancerous growth that would not heal. It started as a small, open area on her neck that would seep, almost like an infected wound, and we speculated that she may have been bitten by a rat.

Generally, wolves have an amazing capacity to heal. In the wild they quickly recover from broken bones, wounds, diseases and infections, and illnesses or injuries that would floor a domestic dog for weeks are shaken off in a trice by wolves. Initially, the wound on Kenai's neck was monitored but dismissed by John, at that time head of wolf welfare. It wasn't until I pointed out that it didn't seem to be improving that we called in the vet, who put her on a course of antibiotics. Days turned to weeks, and still the wound persisted, developing a lump under the seeping hole. Antibiotics made no difference, and the situation wasn't helped by Kenai scratching at it.

Eventually, we isolated her on the yard, and devised a means of preventing her worrying it. An Elizabethan collar – like those dogs wear after an operation – was placed around her neck, with the larger opening facing backward over her shoulders and the wound; held in place by a dog harness. This arrangement precluded Kenai from using her powerful claws to rip open the wound. Unfortunately, this ingenious contraption caused a problem in itself by cutting into Kenai under her forelegs, which meant we then had two areas of injury to treat and keep clean.

The decision was taken to operate. John couldn't get time off work to organise this, so I and Juliette, another senior handler, took Kenai to the local veterinary surgery for the procedure. I remember we were really concerned for her: this had been ongoing for a while, now, and we had no idea what the vet would find once she was anaesthetised. And we were also a little concerned for ourselves, as this was the first time I'd headed up a task like this without John being in attendance. Kenai knew both Juliette and I well, but we had never tested our relationship to this degree.

We said our goodbyes to her that morning, standing on the yard, just before we loaded her up in the trailer. We even took photos – the only ones I have of me with Kenai.

Kenai, trained by Roger, was the ultimate professional, and loaded and travelled easily the short distance to the vet. There, Juliette volunteered to go inside the trailer and hold her while the vet sedated Kenai via the air lock (a system of gates inside the trailer which allow us to safely load and unload the wolves without them escaping). For any medical procedures we liked to knock out the wolves in the trailer, and wake them there, too, thereby minimising any stress.

Kenai went to sleep quickly, and we stretchered her into the operating theatre. Juliette and I had a long and nervous wait – which felt like hours – whilst the vet worked on excising the lump. At some point John arrived to wait with us.

Eventually, Kenai was brought back out. The suture wound was massive, and ran from her ear to her shoulder. Unbeknown to us, the lump we could feel was just the tip of the iceberg. It went a long way into her neck, and turned out to be the size of a large orange. Thought to be malignant, the vet suspected it had probably spread to other parts of her body. We waited for a while to allow Kenai to come round a bit from the anaesthetic, and then set off on the ten-minute journey home. As John had his car and I was not, at the time, comfortable about driving the trailer, it fell to me to travel with Kenai in the back of the trailer, as we did not want to leave her without supervision when she was disorientated and groggy from the drugs. Although muzzled, this was only a cloth type, which I had no doubt she could work around if she felt so inclined!

During the journey, due to the vibration and noise of the trailer, she came to a fair bit more. I didn't want her to thrash about as she regained consciousness, so lay on top of her to hold her still, which she took remarkably well, and we arrived home in one piece, if a little rattled.

Quickly collaring her, we took Kenai on to the hard standing and into a kennel filled with straw. She jumped up onto the bed area and John and I removed the muzzle. I remember thanking her for not biting me, after which we left her alone to sleep off the effects of the anaesthetic. It did not take long; very quickly she had regained her appeite and was bossing us about.

We again rigged up a vet collar over her shoulders to prevent her claws pulling out the stitches. She was a model patient, though, and ten days later stood quietly on a lead for me as the vet removed the stitches. We let her back out into the enclosure with Kody.

We fed Kenai various supplements to boost her immunity, and kept a close eye on the wound site. Within months, though, the lump had begun to return, and we decided not to put her through a second operation. I cleaned the area daily, and Kenai was always keen to come onto the yard for me. I didn't use a collar with her, and rarely had anyone else with me: I found it easier that way with all of the wolves. I could concentrate on reading their signals and knew when I could push forward with a treatment and when to back off and give them a minute to settle themselves. I never once got bitten. I truly believe they knew I was helping them.

As spring turned into the balmy summer of 2006, Kenai's breathing became laboured, and she began to sleep a lot more. It was obvious the cancer had progressed to her lungs, and it was only a matter of time before we lost her. We agonised over the timing. Would she be able to pass over herself or would she need help? In the end we booked the vet to come two days hence. She didn't appear to be in pain, but was on meds just in case.

That evening, lying in bed, exhausted, the phone rang. It was John: Kenai had deteriorated and the vet had been called to put her to sleep. I raced up to the Trust to say my goodbyes. When the time came, though, I opted not to go into the enclosure with John and Colin (Colin had come down to secure Kodiak on the yard so the vet could safely enter the enclosure). The three of them went up to where Kenai was resting in the long grass, whilst I stayed outside and sat on one of the picnic tables. I didn't want to crowd her, believing that, out of everyone, it was John and Colin who knew her best. I sat thinking about her, sending her my love, and telling her it was time to go. I'm not sure what I believe when it comes to the spiritual side of things, but I swear I knew the exact moment she passed over: it felt like a great weight had been lifted from my shoulders, and I felt light and even joyful. I sat and imagined Kenai running over endless prairies, wild and free at last.

We always made sure the wolves saw the bodies of their dead pack mates, so that they understood what had happened to them. I truly believe it helps in the grieving process (oh yes, they do grieve).

Kodiak was led in on a leash by Colin, and he tentatively sniffed Kenai's still body before growling briefly and walking away. We housed him in the holding pen next to Duma and Dakota that night.

Kenai had such presence that, for months after, if Duma and Dakota appeared wary of something we joked that Kenai had come back to haunt them. They might look up at the top of the kennel block, then decide not to come on to the yard. Although we could see or hear nothing that would upset them, I would jokingly tell Kenai to go away, and weirdly, that sometimes seemed to work, and the girls would walk through the gate. It made me laugh as I fondly remembered the old girl. Kodiak quickly moved in with the two sisters, which Kenai would have hated as Kody was 'hers.'

In telling you Kenai's story I hope to portray a sense of her personality, her fortitude in adversity, and her strong spirit. She and Kody were the start of the UK Wolf Conservation Trust, and her passing marked an ending of sorts, severing the link with Trust founder, the late Roger Palmer. I hope they found each other again.

Just hanging out

Down-time was always my favourite part of the day, when the show was over, the crowds had dispersed, and the day was drawing to an end – or hadn't yet properly begun. A time when bonds were formed with both wolf and human friends – laughing, chatting and hanging out – special moments in time with so many memories.

The best part about having other handlers around was it allowed all kinds of fun opportunities, such as going into the enclosure, or walking the wolves without the stress of having the public around. My favourite walks were the early morning enrichment marches we did with the Euros. I call them marches because they were usually done at top speed, as the Euros loved to stride out, fling themselves at concealed prey, scent roll, and generally try to outwit us.

Being fussy about who they hung out with meant there were no trainees, just us experienced handlers: stalwarts who had known each other for years, and were all really good with the wolves – though still managing to have a laugh. It didn't matter if it was minus six, or lovely, early morning summer sunshine, we would be there in advance to enjoy our time with Alba, Lunca, Latea, and the other wolves.

Of all the years I worked with the wolves, 2006 was one of the hardest but also the most notable. We lost Kenai, and were also still nursing Alba from a terrible injury (more about which in the next chapter). But we also gained the cubs Torak, Mosi, Mai, and Mika, and hanging out with those little guys was always an unforgettable experience.

Alba's story

Alba was the most powerful and majestic animal I have ever had the pleasure of knowing. It wasn't just his physical size that impressed, but his presence, also, although I never felt threatened by him.

Although I was definitely in Alba's fan club, others had seen a darker side to his nature, and he had no qualms about 'taking handlers off his Christmas card list,' as one particular handler used to say. When, on the night of 25 June, 2005, Alba suffered a life-changing injury, it was devastating to see this giant of a wolf reduced to helplessness.

I first heard about Alba's accident at 5am, the morning of 26 June that year, when I arrived at the Trust. We were taking some of the wolves to an event in Somerset, and show days always meant an extra-early start, as it took us around two hours to erect the travelling enclosure that the wolves were displayed in. I pulled in to the car park at the same time as handlers Dom and Wendy, who, I could see, were clearly upset: it was they who had found Alba, late the previous night, when they had come to feed the wolves, and put them to bed.

We preferred to leave the wolves out as late as possible, so it was usual to feed at around 10pm. Out of all the wolves, you could guarantee that Alba would be waiting at the gate, demanding his food, so when he wasn't there, and didn't appear, Dom and Wendy went into the enclosure to look for him, and discovered him collapsed and in considerable pain. It seemed that, sometime between 7.30pm and 10pm that evening, something catastrophic had happened to Alba.

The vet came immediately, and diagnosed a spinal injury. Although euthanasia was suggested, the decision was made to give Alba a few days' grace for further observation and consultation with leading experts in spinal injury. He was sedated and injected with painkiller. That Alba was a socialised wolf saved his life that night: if he had not been hand-reared, there really would have been only one option.

Wendy, Dom and John (head of wolf welfare at the time) managed to get Alba onto the hard standing with the help of a duvet, and someone stayed with him overnight. Alba was in shock, and had only limited body movement. Worst affected was his left foreleg, which had no movement at all from the shoulder down, and hung from his body like a piece of meat. In addition, there were problems with his left hind leg, as well as an obvious injury to the left-hand side of his face and eye.

My background in animal welfare told me there were numerous options we could try, and I also had a list as long as your arm of friends and colleagues who dealt with this type of injury on a daily basis. For the time being, though, I was needed at the event, and I left the Trust with a head and heart full of emotion. It was tough that day, carrying on, knowing what was going on back at the centre.

During the day a little of the puzzle was pieced together about what may have caused Alba's injuries. Alba and his sisters, Latea and Lunca, had a habit of chasing each other around the enclosure at full speed (at speeds of between 25 and 45mph), often not looking where they were going, especially if playing tag. When the enclosure was built a few years previously, many trees had been planted, and staked to help the sapling grow straight. Wolves are naturally destructive, so everything within the enclosure was substantial, but one tree stake (at least 15 centimetres (about 6 inches) thick) was found snapped in two. What it must have taken to do this, I simply can't imagine, but we could see what resulted from it: Alba was paralysed.

When we arrived back that evening the vet was there. More painkillers and sedation were given, and, as there had been some deterioration, euthanasia was again considered and rejected. An opiate-based injection was left in case we needed it during the night. John had also given Alba the homoeopathic remedy of arnica at 30c (for bruising, pain, and shock, but which can also stimulate healing of damaged tissue), and hypericum at 6c (for nerve pain after injury, head injury, and shooting pain) every half hour. They had also been trying to get an electrolyte solution into him to prevent dehydration, and help with shock.

I volunteered to stay with Alba that night. I was the only one trained to give subcutaneous (under the skin) injections, and I knew that the Tellington TTouch I am trained in would help keep Alba calm and comfortable.

Nikki said she'd say with me, so, after a quick dash home for sleeping bags and provisions, we were soon back.

Alba hadn't eaten since the night before last, so I tried him on some cat food I'd brought from home, which he ate a small amount of. Getting food and fluids into him was clearly going to be an issue, and over the next few days we came up with ingenious ways to do this.

That night I lay beside Alba with one of his paws in my hand, so that I could tell in an instant if he tried to move. Periodically, he would begin to tremble, and I would sit up and do some more TTouch body work on him until the trembling subsided and he settled. Nikki and I didn't sleep; just lay there with him: watching his every move, praying for some positive signs but knowing that the outlook was bleak.

It was a godsend that it was a warm, dry summer. Alba had only a straw bed between him and the concrete of the hard standing, the area between the enclosure and the kennels, where he was being temporarily housed. The kennels were much too narrow inside for him to go there, but he was dry at least. It was a long night.

Towards dawn we noticed that Alba was becoming more distressed: his trembling increased and he became agitated. At 5am we rang John to confer, and the decision was taken that I should inject Alba with the opiate that the vet had provided. A subcutaneous injection goes just under the skin but not into the muscle, and the obvious and usual place to administer this is in the scruff of the neck, where there is an excess of skin. In Alba's case, however, I felt this was inappropriate due to the trauma in this area (he could barely lift his head). Given Alba's feisty reputation, coupled with his being in pain, injecting him safely without getting bitten was not going to be an easy task.

Nikki and I exchanged looks. I decided that an alternative injection site would be a fold of skin between the right hind leg and the abdominal area, if I could lift a handful of loose skin into a pocket to inject the drug. Nikki reached for a towel to cover Alba's eyes, the one in the most danger, closer to his head and those powerful jaws. I had one chance at this and I had to do it quickly. We counted to three, the towel was placed over his head and I quickly injected. It was over in seconds with no obvious reaction from Alba. We watched and waited until his shaking subsided and he rested peacefully once more. Our emotions were shot: we'd been up over 24 hours but neither of us felt we could sleep.

By the time John arrived with breakfast for Nikki and me I was in bits, due to a combination of sleep deprivation, lack of food, and stress. The three of us discussed the next steps: a pattern which continued over the following few weeks. I'd suggest treatments, people to call for help, and provide the main nursing care. John would okay or veto my ideas, and make the final decisions, and Nikki would organise the rota so that Alba had people with him 24/7.

We all had times of hitting rock bottom, in tears but supporting each other. Luckily, if one of us was down, another would be more upbeat that day, and pull through the rest. We also very quickly came to trust Alba to help us make the right choices, relying on his strength of character, the look in his eye, and his physical strength and determination. If Alba wasn't giving up, then neither were we.

At 9am the next day, 27 June, I was on the phone to my friend, James French, a reiki master and animal communicator, asking for his help, and he was the first of an amazing group of people who gave up their time to help Alba, often without charge. We arranged for James to send distance healing the next day when I could be with Alba to watch his reaction. We also had a meeting with the senior partner at the veterinary practice, who advised us that a significant improvement was essential over the next few days if Alba was to survive. He was still in shock, so x-raying him wasn't feasible, as the necessary anaesthetic would be a risk. The journey to the vet could also cause additional trauma, with associated risk of aggravating the injury. Whatever happened the treatment was the same: rest and pain relief. Various sources advised we were looking at an eight-week recovery period, with the prospect that Alba may never walk again a distinct possibility.

That day we conferred with experts about additional homoeopathic remedies to use alongside conventional pain relief. Use of arnica and hypericum would continue; later we would add other remedies.

We were having problems getting painkillers into Alba. Understandably, nobody wanted to put their hand down the back of his throat when his neck was obviously injured and his pain level high. It was suggested we crush the tablets and put them in a syringe with the top cut off, inserting the syringe into the side of Alba's mouth as far back as possible. A sports drink bottle was used to squirt water into Alba's mouth, and, hopefully wash down the meds. Not a great way of doing it but one which worked well enough. As he couldn't lift his head much, using the drink bottle to squirt fluid into his mouth was the best way to keep him hydrated. Alba learned quickly, and when we offered water he would open his mouth to receive it.

We began to worry that Alba was overheating, as the days were hot, so rigged a tarpaulin over the top of the yard, and added wicker fencing panels to the chain link sides of the yard to provide cooling shade. The wolves generally didn't like this type of arrangement when used at shows, and preferred to lay out in the sun, but we felt it was necessary as, of course, Alba couldn't move. At some point it was bound to rain, and this would provide cover.

Since his accident two days previously Alba had been very quiet, neither howling or vocalising: he was hiding out; feeling vulnerable. He was also lying on the side most affected by the accident – his left. We kept him as clean as possible when he urinated, to prevent scalding from the urine, and I continued to do several daily sessions of Tellington TTouch on him. These always seemed to calm him, and he would usually fall asleep whilst I worked.

He still refused to eat, however, which worried me a lot. Wolves can go for a long time without food by living off their reserves, but we wanted to provide nutritional support to help Alba heal.

One particular day – Black Monday – was probably the worst for many of us at the Trust, as we wondered whether Alba would have the strength and fight to pull through. He was so shut down and still in considerable pain, the tremors always returning when he was due his next pain relief. And we constantly questioned whether we were doing the right thing for him ...

Alba's sisters kept a constant vigil over him and us, hanging out just the other side of the fence. They howled a lot and became agitated, pacing up and down. They even – clever girls – destroyed the tree that the broken stake had supported, chewing it right off. Confirmation of what actually happened, perhaps?

I'd been in touch with Gavin Schofield, a very talented osteopath who runs a busy animal practice. As luck would have it, Gavin was in our area, and could see Alba that night. As Alba does not 'do' strangers, it was a mark of his understanding that he needed help – and Gavin's skill and approach – that Alba allowed himself to be treated. Gavin determined that Alba's C6 vertebra was extended and compressed, resulting in muscle spasms in the neck, which I'd certainly seen, and felt a tightness in that general area.

Gavin also worked on Alba cranially. We had noted that Alba guarded the left side of his head and face, and the left eye appeared damaged. Gavin confirmed that this must have been the impact site, and worked to release the compression there. I had to be at work when Gavin came but John was impressed with his work, and afterward Alba had so much more mobility in his neck he was able to lift it higher.

Still receiving 24-hour intensive care, that night Alba was quieter and rested better than he previously had.

28 June (Tuesday)
I was back with Alba at 7am on the third day, and gave him another TTouch and stimulation session. For the first time I felt a tiny movement in his left hind leg and paw. I cannot describe the flood of emotion I felt. Alba's life still hung in the balance, but any sign of recovery was significant and very, very welcome. That day he seemed so much more alert, beginning to take an interest and look around, and even had a scratch reflex from the right hind leg when I worked around his neck. The shock was definitely subsiding.

I managed to get Alba to take a little pureed cat food mixed with gravy and electrolytes. Whatever we offered him in the form of proper food, such as slivers of beef or mince, he refused, but who knew what his head and jaw felt like after the force of the impact? I was really concerned that he was lying in urine, so we lifted him up on a towel to clean him and replace the bedding. Where we were housing him was far from ideal, and I was concerned about bedsores developing. I racked my brains – and asked some animal colleagues – for what we could use that would cushion him but also be wolf-proof, but had not yet come up with anything feasible.

I did two TTouch sessions on him that morning. His neck was definitely spasming less that day, although he was still very reluctant to let me touch his head and face on the left side. He really responded to my working on both fore and hind left legs, lifting his head to see what I was doing. I noticed spasms throughout his whole body often as I worked. He even extended his left hind leg and pushed really strongly against my hand, which he'd not done before.

As mentioned, I'd arranged for reiki master and animal communicator James French to send Alba some distance healing and communication that day, and, at the allotted time, I sat quietly in the corner of Alba's yard and watched him. I've had reiki myself – in person and by distance – and with me it can throw up some really weird sensations. Alba made all sorts of movements during the 30-minute session, and at one point even tried to get up: really putting all his effort into it, getting his neck and shoulders off the ground (this worried me until James later told me he had asked Alba to try and stand).

By the end of the session Alba's breathing was calmer, deeper and steadier. He had lapsed into a deep sleep, and didn't even stir when I got up to leave.

I left him for an hour and then returned to do some more TTouch body work. It is really important with major injuries and traumas to, if possible, work little and often. The combination of pain relief, homoeopathy, osteopathy, reiki, and TTouch had allowed Alba's neck muscle spasms to subside, and he also pulled away from me when I worked in-between the toes of his left hind foot: an indication he was regaining feeling in that leg.

It was, however, difficult to work on his head, as he was still a little reluctant for me to touch this area, and his left foreleg still gave cause for concern due to the lack of movement. Even so, Alba showed remarkable tolerance in allowing us to do so much with him, and I was beginning to trust that, despite the fact that his pain and distress could cause him to inflict serious damage, I could do just about anything with him. He took all help with good grace.

At lunchtime I talked to Angela Griffiths, a friend and colleague at Greyfriars Hydrotherapy and Spinal Rehabilitation Centre in Guildford. She told me that it would take at least two weeks for Alba's bruising and swelling to subside, and was keen for us to check if there was any nerve damage or disc rupture. Although significant recovery could be expected by eight weeks, further improvement was possible for up to eighteen months, post-injury. Physiotherapy would be the next step. Christelle, Greyfriars' physiotherapist, agreed to contact our vet to arrange permission for this, and gather the relevant information.

Although Alba had taken limited amounts of cat food, I was keen to get him to eat something besides this. We used hotdogs as treats – they were a good motivator as the wolves loved them – so I thought there was no harm in trying these.

The reaction was astonishing. We don't generally hand-feed wolves, for obvious reasons, but knowing Alba's limited movement and lack of appetite, I thought I'd show him one. Immediately he lifted a third of his body off the ground and snapped off most of the dog, just millimetres from my fingertips. I offered another; he took it again. At one point his teeth connected with my fingers but that wolf, even with a life-threatening injury, knew he'd made contact with my skin and not the sausage, and immediately let go. I had no mark, no bruising, no pain, yet he could have easily bitten off my finger. Even in his condition, he had enough self-control and bite inhibition to avoid injuring me: this wolf was something else! My love and respect for him just grew and grew. Alba ate five hotdogs before deciding he'd had enough.

Before I left to go to work that afternoon I did another ten minutes of Tellington TTouch body work on him. He was visibly relaxing now: he had tended to hold his right hind leg rigidly tucked up tight to his body, but in this session he released and stretched it. In fact, he had less tension and guarding in his whole body.

After a busy morning we mostly left Alba in peace for the afternoon, just going in to give him his homoeopathic remedies every two hours, and offer water and food. In the evening he ate a lot more hotdogs – sixteen in one sitting! There was a storm that night, so the decision was made to move him into the kennel for added protection from the elements, where he was tucked up safe and dry with the door open. When he was checked just after midnight, he was back out in the yard, having managed to somehow drag himself there, the first time he had been able to move any distance. The handlers struggled all night to keep Alba dry and the tarpaulin roof free of water, and dry straw had to be laid several times. At about 1am he ate a further 14 hotdogs, though still refused any mince or slivers of beef.

29 June (Wednesday)
At 6am the next morning, Alba was curled up and bearing weight on his left elbow. Previously, he'd been laying flat; unable to lift his head or shoulders much. He was relaxed, alert and looking around, pleased to see us. There was more life in his eyes, too, though he was slightly cold and trembling, which could have been because he was due pain relief in the next couple of hours. I cleaned the yard area and laid fresh straw, with Alba looking on.

At 8am Alba's pain meds were given and water offered, and I gave him another TTouch session. He had so much more movement throughout his whole body, and was beginning to find ways of getting about: kind of sliding his chin along the ground and using his hind legs to push. He was very determined about it. His lower left foreleg still looked very odd, and had little movement, but he was stretching the whole leg from the shoulder, and could obviously feel sensation in it, especially when I worked on the foot and lower limb. At last, it looked more like living flesh than meat, and, in fact, both left legs looked more as they should.

Whilst working on him I noticed a swelling at the base of his neck, just at the right-hand side of the spine, and he was guarding his right hip. After the session, I offered some more food but he wasn't interested – still full up of hotdogs, I guess. He curled up to sleep, again bearing weight on the left elbow.

Alba slept until 11am, and I worked on him again when he woke; this time eliciting a very interesting – and different – response. He had a very tight area running down the front of his right shoulder blade, and when I worked here I saw his whole body spasm, and he tried to move. I'd noted several times that day that his left leg was bent at the carpal joint, and during this session I actually watched him straighten it. The rush of joy that ran through me was electric: he'd moved that seemingly inert piece of meat! He could feel me working on it and he was moving it. We all ran around jumping for joy, in celebratory mood.

During the afternoon he had another distance reiki session. The healing seemed to be directed to the mid-back, right hip and left shoulder areas, and, at one point, Alba tried hard to stand. He got his front legs under him, but succeeded only in lifting himself on to his carpal joints, whilst swinging around and biting at and licking his right hip. As seemed to be the pattern, though, by the end of the session he relaxed and fell fast asleep.

We decided to add Rescue Remedy and a multivitamin supplement to his electrolyte solution to further support his body, and he took some of this at lunchtime. More hotdogs, clean bedding and a clean-up session followed, after which he spent the afternoon sleeping.

A team stayed with him until midnight, but we felt comfortable enough about his condition to leave him alone overnight for the first time. John and I would come in at 6am to check on him.

30 June (Thursday)

It rained overnight and Alba got wet. His coat, this time of year, had only one layer – the long guard hair – with no thick, warming undercoat. I dried him off with a towel and replaced some of the bedding. His lower left leg again appeared disconnected from his body: it obviously needed more stimulation. By touching the area in non-habitual ways through physically moving the skin and joints, it's possible to stimulate the neural pathways and retrain the body to work as it should, which is why it's important that regular treatment is given.

After his 8am meds I gave the yard a really good clean: Alba had voided his bladder and bowels during the night, and I was concerned about hygiene. He seemed to really enjoy the stimulation of my doing this, and helped me spread around the new bedding, moving about a lot and tunnelling in the fresh straw by pushing himself along on his chin, powering with the hind legs and sometimes the right foreleg. His left foreleg was getting stuck underneath him in a really dangerous position, but he worked out a way of flipping over on his back to release it and then repeat the exercise: extremely scary to watch as I thought he'd break the leg. Occasionally, I had to help him release the trapped left leg, then off he'd go again. John rang me whilst this was going on, and at one point I had to throw down the phone and run to Alba's rescue. Actually, he didn't really need me, but it took a few major heart palpitations on my part to trust he could extricate himself.

Once Alba learnt this slide and flip technique, there was no stopping him; he was everywhere, getting close to the fence to interact with his sisters, who often lay just the other side, or following me whilst I was cleaning, and generally scaring the hell out of me in the process. This wasn't what the vet meant by rest, I felt sure!

His TTouch session that morning was for general relaxation, as Alba looked a little sore from all his earlier activity. During the day we consulted with the vet. Although we had seen a lot of improvement, there was still a long way to go, and concern was growing that we'd missed something because of the continuing issues with his left hip and left hindleg. We needed an accurate diagnosis: it was time to take Alba to the veterinary surgery for x-rays, which we arranged for the following morning. It was make or break again; we were all nervous. If the results were bad, would we be saying goodbye to him in less than 24 hours?

After his noon meds Alba had another distance reiki session. There was lots of lip-smacking and tremors to begin with, but after five minutes he fell into a truly deep sleep, and remained that way for the rest of the session, not even waking when I left. James said he could feel that Alba's left shoulder was drawing a lot of healing.

That evening, when the team offered food, not only did Alba eat a can-and-a-half of hotdogs, but also about a pound-and-a-half (680g) of beef and some chicken. Finally, he was taking proper food, which should better prepare him for the general anaesthetic the next day. And he was really trying to move, licking his hindquarters, and at one point managing to get about, weight-bearing on the left leg with it in the correct position. He was very bright and alert. During sleep, all four of his legs twitched as he dreamt.

1 July (Friday)

6.30 the next morning Alba was bright and alert, the spasms throughout his body a regular occurrence now, in anticipation of his imminent pain medication. After checking the other wolves and attending to morning routines, I returned to Alba at eight to find him in the kennel, his left leg stuck under him in an uncomfortable-looking position. He was growling and his eyes were hard: he was obviously in pain.

Although no sane person would approach a huge, growling wolf without expecting to come off worse, I cautiously edged toward him from behind. It wasn't easy to approach safely due to the limited space in the kennel area, most of which was taken up by raised beds, with just a narrow corridor running along the front. A combination of my nudging the leg and Alba trying to move – air-snapping at me the entire time – resulted in him managing to achieve a more comfortable position.

After five minutes of TTouch to calm both of us, Alba was asleep and breathing deeply. I'd never been so glad to be in Alba's club; thank goodness he trusted me. Those not on his list of personal favourites were tolerated when nursing him, but we were under no illusion: even in this condition he could cause someone serious damage.

His x-ray was booked for 12.30pm. We'd gathered together a team to transport him to the surgery, but had no idea how he would react. The duvet he had originally been carried in on from the enclosure the night of the accident was laid next to him, and the boys carefully rolled him onto it, lifting him and walking the short distant to the converted livestock trailer that the wolves were often transported in. Alba was used to this, therefore, but the jolting was bound to cause him pain, so it wasn't the best scenario. Some of us – I don't remember who – travelled in the back with him. Surprisingly, Alba took it all in his stride: all-in-all a model patient throughout the majority of his recovery.

Arriving at the practice, we dropped the tailgate and the vet injected Alba with the anaesthetic, via his leg. Alba didn't growl or struggle, and was out in moments. A stretcher was used to convey him to the surgery and awaiting x-ray machine.

We had a nervous wait. I was in the process of quitting smoking – not the best time to do it – and had my very last cigarette, nervously waiting outside, trying to peek in the window; praying that the outcome would be good. I vowed I'd never smoke again if it was. At this point, six days after the accident, we were all exhausted and running on adrenaline. None of us had slept much, eaten a decent meal, or spent time at home in that period. I'd even fallen asleep at work one afternoon, sitting at a table with a cup of tea!

At last John and I were called in to look at the x-rays – Alba still asleep on the table – which showed a stable fracture to C2 vertebrae in the neck. As the fracture had not severed the spinal cord the prognosis was guarded but better. It seems Alba's extremely strong neck muscles had helped keep the fracture stable.

When hunting, wolves can get thrown about by large prey, so the neck needs to be strong; this simple fact of anatomy probably saved Alba's life – that and being able to care for him as he was socialised. The vet advised it could take up to eight weeks for Alba to walk again – if he ever did – and even if he could stand it was unlikely we'd see a complete recovery. Alba would be disabled but would survive if he could manage to stand.

After long and heartfelt discussions, because Alba was so fit and had shown such progress and determination in the last week, it was decided to continue with his rehabilitation. Another stay of execution: a good day.

That afternoon he made a good recovery from the general anaesthetic, and continued moving around. In the evening he ate more beef, took his meds, and had another short TTouch session. After final checks and meds at around half twelve, he took himself off into the kennel to sleep. We all slept well that night.

2 July (Saturday): one week since the accident

At 6.30 the next day when I checked on him, Alba was still in the kennel and looked comfortable. I gave him a short TTouch session, his homoeopathic remedies, and a small amount of food and water.

After his 8am pain relief I did a photonic torch session on him. This piece of equipment directs a special light on particular areas of the skin, and promotes healing by increasing electrical conductivity, stimulating the brain to release chemicals and hormones that relieve pain, increase immunity, and encourage healing. It works as acupuncture does by stimulating pressure points and triggering nerve messages. Often in spinal nerve damage the brain 'forgets' the injured area after a while, as a damage limitation exercise. The torch reawakens the connection between the tissues and the brain.

Since Alba's accident I'd been receiving support and advice from a wide range of professional friends, and Sharon King, a fellow Tellington TTouch practitioner, kindly offered to lend me her torch to try.

Laying out in the kennel corridor, looking out of the open door, Alba was, at first, very interested as I followed the instructions I'd been given, sitting on the raised bed above, leaning over him. The reaction was amazing: not only did he fall asleep halfway through the treatment, he didn't even stir when I got up to leave. Normally, it's not possible to move around a wolf even slightly without their being completely aware of every move you make ... but Alba slept on.

I'd noticed that Alba was beginning to develop pressure sores on his left shoulder (and later on the carpal joints (knuckles) as he dragged himself around). Nobody else seemed to take this seriously but I knew they were bad news: bedsores can actually kill if nothing is done about them, as bone becomes exposed and infection sets in. I racked my brains and talked to loads of people about how best to treat these, and, in the meantime, we tried putting matting under him in an effort to cushion his joints against the hard, cold concrete. This wasn't very wolf-proof, unfortunately, and was chewed quite a bit.

We kept him quiet for the rest of the day. Now that the shock and trauma of the actual accident had subsided, sleep was the best healer, and we all needed some, not just Alba. He continued to take food, showing a preference for chicken carcass, perhaps needing the calcium from the bones to heal. He was still on nineteen-hour daily care, on his own from just 1am to 6am.

3 July (Sunday)

By this point, we had begun to wean Alba off the intensive care regime, recognising he needed sleep and downtime, with help nearby if required. He was beginning to prefer being inside the kennel and not on the yard; 'denning up' due to feeling vulnerable, possibly. We had noticed that he didn't join in with his sisters during howling sessions. Maybe he stayed silent to protect himself from possible threat? Top-ranking wolves have distinctive howls, and it could be that Alba didn't want anyone to know of his weakened state.

Moving about, he was perfecting his unique slide and roll technique, though now pulling his back legs under him and pushing himself forward, on to his right foreleg, but knuckling (walking on the knuckles, rather than the pads of the foot). He was able to move around a lot in this way, though had to stop and roll often to release the left foreleg which didn't move, and became trapped underneath him.

The big change that day was that he was beginning to lay in a more upright position: sphinx-like with his hind legs under him and front legs out in front, though this was sometimes more half-sphinx with his hind legs to one side. The pressure sore on his left shoulder had begun to open up, though, which was a real worry. I'd taken some time off over the preceding weekend, when there had been more people around to care for him, but it felt as if I'd taken my eye off the ball. I'd still not come up with a suitable alternative to straw with which to cushion him, keep him warm, *and* be wolf-proof. We applied some sudocrem to the sore in the hope that this would help.

Lunca and Latea, his sisters, continued to keep a close eye on everything we did, and Alba often interacted with them through the fencing between the yard and enclosure. Once, when Alba was much less mobile, we'd let the girls onto the yard, and they'd been really careful with him, though he clearly felt the need to muzzle-hold and

Continued page 81

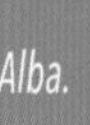

Alba.

Duma.

Mai. (l)

Dakota. (r)

Mika. (r)

Mosi.

Kenai. (l)

Lunca. (r)

Torak.

Latea. (l)

Kenai's wound before her operation. (r)

Kenai two days after the operation, wearing her adapted veterinary collar.

Duma.

Torak with his favourite: pumpkin.

Torak, Mosi and Mai playing in the snow.

Dakota. (l)

Mosi. (r)

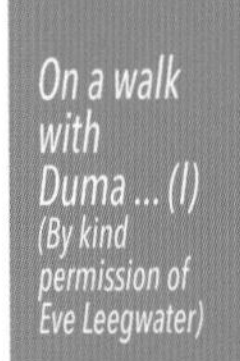

On a walk with Duma ... (l) (By kind permission of Eve Leegwater)

... and hanging out with Dakota in the enclosure. (r)

Alba resting after his first swim. The damage to his left eye is evident. (By kind permission of Greyfriars Rehabilitation Centre)

John swimming Alba at Greyfriars Rehab Centre near Guildford, England. (l)

Alba after his first swim: clearly tired after just a few minutes. He improved greatly over the coming months. (r) (By kind permission of Greyfriars Rehabilitation Centre)

The cubs arrive at the Trust, 8 May 2006 ... (l)

... and a few weeks later, at the end of May. (r)

Torak playing with my shoes ... (l)

... and a warm and sleepy pile of pups. (r)

6 June 2006: first real trip out of the 'den.'

June 2006

Dominance role-play.

Mika selfie, curled up on my lap. (l)

Mika sleeping soundly ... (r)

Torak in thoughtful mood ... (l)

... and being cute!

Mai showing early hunting behaviour, pouncing on mice. It wasn't long before they were all catching them.

As with all youngsters, water held a particular fascination.

Squabbling over a bone.

Still playing and sleeping in a puppy pile.

July 2006

Feeding time.

To get the cubs used to the water bucket and the noises it made, I would tap the sides as I brought it to them. (l)

August 2006

Trailer training – for Mai, especially. (r)

Watching for trouble: Mai guards her food. (l)

Torak wants a kiss from cameraman Ian Moss during filming of Ambassadors of the Wild. (r)

All grown up, now: Torak, Mosi, and Mai playing in the snow.

Io from Wildwood Trust, Kent. All of the Wildwood wolves are related to Alba, Lunca, and Latea in some way. (r)

Apollo, father to Alba, Lunca, and Latea shortly before his death at the grand old age of sixteen. (m)
(Wildwood Trust, Kent)

Wildwood's Nadja; a shot I took when visiting to see the wolves and catch up with my friend, and Wildwood's Education Officer, Anne Riddell.

Cuddles from Tatra at Paradise Wildlife Park with my good friend Pia Gismondi. (l)

Tatra and Misha from Paradise Wildlife Park, are both socialised, hand-reared wolves. (r)

Peyto (l) and Kaya from the Anglian Wolf Society. Related to Alba, Lunca, and Latea, sadly, Peyto is no longer with us.

snap at them to protect himself from any pain they might inadvertently cause. The sisters had then become a little agitated with the handlers, perhaps perceiving us as a threat.

It's so important that pack members aren't separated for long. Doing so causes them intense emotional strain, and even sometimes results in a pack member not being allowed back into the group. If there had been another male in the pack we probably would never have been able to re-integrate him, but as Alba's sisters were devoted to him, we decided to let them on to the yard again. It was lovely to see the three reunited, and although we still had to be careful of the girls' behaviour towards us, they had a great time together. There was no aggression, and Alba loved the contact and became really lively. It lifted everyone's spirits – human and wolf.

It had been a long and worrying week, but were we over the worst? After what we knew would be difficult forthcoming months, would our special boy stand again? And if he did, would he be able to function in the fast-moving dynamics of pack life? Would the girls fully accept him again, and would he be able to control the tumultuous relationship between Lunca and Latea? So many questions about the future. First, though, we had to get through the second week of rehabilitation.

4 July (Monday)

Alba seemed fairly comfortable this morning, a little tight around his shoulders and neck, maybe, and the usual morning tremors, but ten minutes of TTouch relaxed him and he fell asleep.

His pressure sore showed signs of necrotic tissue today, so the vet prescribed antibiotics, which we added to his crushed painkillers, using the syringe technique to get as much of the powder down him as we could. He was so good about it: it must have tasted and felt awful in his mouth and throat. Although we tried to quickly wash it down with water, the poor lad didn't look very happy about the twice-daily ordeal.

The vet also gave us a special gel called Intrasite for the worst sore, and dressings to cover it with. John and I tried to clean and dress his left shoulder, but Alba wouldn't play ball. It didn't help, either, that he was inside the narrow kennel, as each time we had the area clean and the dressing ready Alba would roll back down and lay on the shoulder we were trying to treat. And he was growling at us a lot, too.

Very soon we were covered in the gel, with numerous dressings strewn about the place. The one good thing that came out of this debacle was Alba pushing from and bending his left foreleg: he was obviously regaining more and more function and feeling in it. As for the dressing, we admitted defeat and went for a cuppa to devise Plan B.

Later, I gave Alba another photonic torch session, which he seemed to love. By the end of the session he had stopped trembling and fallen asleep, snoring softly as I crept out.

We ordered some more homoeopathic remedies – a mix of arnica, hypericum, and symphytum

– adding calcarea phosphorica (calc phos: a homeopathic form of calcium) and calcarea fluorata (calc fluor: calcium phosphate) to help with bones, and the spine in particular. These came in crystal form so were easier to administer, which we did every four hours, now, instead of every two.

5 July (Tuesday)

Alba was calm and relaxed this morning; he had a little food first thing and took his meds well at 8am. The new remedies had arrived so I also gave him those in doses of 12c calc fluor, 12c calc phos, 30c symphytum (again, great for bone healing), plus the original 30c arnica (for bruising and swelling), and 30c hypericum (for nerve damage).

During the morning Alba rolled on to his right side, one of the few times he had exposed his injured side. He didn't stay in this position for long, but I was able to check his pressure sore, which was dry and clean. We'd been advised not to try and make him lie on his good side as he would prefer to guard the hurt area. Because of how he got around – and the fact that he was doing this more – sores were beginning to appear on his knuckles, and, similarly, a slight abrasion on the left hip, too. We desperately needed to find a way of cushioning him to protect his limbs as he moved.

Late that morning I had a lightbulb moment – wood shavings, like those used as bedding for horses – and we laid a good six-inch layer of shavings under a top layer of straw in the kennel and out in the run.

The morning passed in the usual routine of cleaning Alba and surrounding area, and doing some TTouch on him. He had progressed to going outside to toilet, which I appreciated as it made it much easier to keep him and his housing clean. For the rest of the day we left our increasingly independent and now slightly grumpy boy to his own devices, only going in to feed and water him, and administer treatments.

6 July (Wednesday)

Alba looked tired this morning. I touched his right foreleg as I was arranging his bedding, and he yelped and snarled at me, so I decided to give him his painkillers a little earlier than usual as he was obviously sore. There was no way I was going anywhere near his head, the mood he was in, so I hid the meds in some chicken which, thank goodness, he ate.

After eating he pulled himself out into the run and back again, apparently better for having moved: it could be he had become stiff overnight. While he was up I took the opportunity to check his pressure sore which was still dry, clean, and clear of infection. It looked to be healing – the wood shavings must be working – what a relief!

Once he had worked out the kinks and the meds kicked in I did some more TTouch on him, mainly on the right hindleg as it was this limb he kept offering me. I'm not sure whether this was because this was a less sensitive area from his point of view, or if he really did need work on the leg, but I was happy to go along with what he wanted.

Later that morning I gave him

another photonic torch session, and, as usual, he was asleep by the end of it, very content.

As had become the pattern now, Alba was left in peace to rest for much of the afternoon.

In the evening, Alba became trapped in the kennel, casting himself like horses sometimes do, his legs trapped underneath him against a wall, preventing him from moving. It took a little time to reposition and free him, because he was snarling and snapping in fear and frustration. Not one of Alba's better days, though it did reinforce the requirement for near-constant care.

7 July (Thursday)

A big day today. Using his chin as a lever, Alba sat up for the first time, motivated by the extra food I'd left on the raised bed. I was concerned he had put pressure on his neck to achieve this, but at the same time delighted that he had sat up – and I mean a proper sit – and we celebrated again. This new development did necessitate a rethink, however, as we had been leaving all sorts of equipment, such as cotton wool and wound-cleaning paraphernalia, in the kennel for easy access, which was now within Alba's reach. Each day he was laying more and more in a sphinx- and half-sphinx-like position, which was lovely to see.

Lunca and Latea had another visit with him on the yard that evening, Alba squeaking and wagging his tail, a huge, wolfie grin on his face. These visits really did cheer up everyone, and were vital to the pack's morale.

8 July (Friday)

Alba's first physio session; Christelle was great with him. She noted that his joints were beginning to lock up (becoming high-toned), which exercises would help to rectify, and also advised that we should begin getting him up on his feet into a standing position. We were to continue Alba's TTouch sessions, and add in some massage, together with shoulder blade circles, shoulder joint, elbow, carpal, and forefoot extensions and flexions, hip, knee, hock and hind foot extensions and flexions, and tail traction. When we stood him we were to gently shift Alba side-to-side and forward and back. His muscle tone was hugely diminished, and we needed to retrain his neural pathways, too.

The exercises that Christelle advised were the next step in Alba's rehabilitation, and we were hugely excited. The body work aspect of the exercises fell to me to do as I had more experience in this area, but getting him up on his feet would be a team effort, requiring a lot of us to manage it safely. Alba was also really heavy, and we obviously did not want to drop him.

Lastly, Christelle left us a laser machine to use on Alba's pressure sores to promote healing.

9 July (Saturday) and 10 July (Sunday): two weeks after the accident

Over the weekend we continued to get Alba up on his feet, which took two strong people. He could weight-bear on his hind legs for a few seconds, but tended to knuckle on the front limbs. We tried to position his legs in the correct stance, but he could stand for only very short periods of time. A couple of times when standing, it seemed as though Alba wanted to take a step forward. Lasering his pressure sores certainly helped with the healing process, though this would take a long time with the left shoulder.

Thankfully, he even began lapping water from a bowl: a great relief as it meant he could keep himself hydrated. We placed several bowls around the area so water was always easily within reach.

11 July (Monday)

Alba was really active this morning. He wasn't interested in the water we offered, as he was now drinking for himself. We stood him up and all four legs locked into the correct position, and he was able to stand unaided for approximately ten to twenty seconds with just a hand supporting his right hip to prevent him falling sideways. This was very encouraging, and we were happy that all the work we did with Alba over the weekend hadn't, apparently, tired him too much.

I wasn't able to do much physio as Alba warned me off with lip curls and stares – he was obviously sore and in need of his morning pain relief. I was able to give him physio on his right hindleg, and a little on his right foreleg, though only part of the left fore- and hindleg as these were underneath him.

The TTouch session was much more successful, however, and he settled down to let me work all over on him. I tried the physio again after, and was much better able to do this, now he'd relaxed. Therefater, this became the routine: TTouch first, after which he would accept the physio, and I could get him to achieve greater movement. The two modalities really complemented each other.

12 July (Tuesday)

A really active day for Alba. Playing in his bed as I tried to clean up and rearrange it, he made me laugh out loud, he was being such a clown. Later that morning, after a really successful TTouch session, I was able to do a lot of fantastic physio, with Alba in a sitting position. He was beginning to weight-bear on his left foreleg, and, at one stage, was lying squarely, all four legs in the right positions, as he pushed up on his forelegs. I could tell that Alba really enjoyed this session: there were no yelps, lip curls, growls or air snaps; he really seemed to participate fully, to help himself. He definitely looked happy and relaxed and more mobile by the time we'd finished.

The only downside that day was that the pressure sore on his left shoulder appeared slightly infected again. I cleaned this thoroughly with a diluted veterinary solution called Hibiscrub, and applied Intrasite gel (an amorphous hydrogel which gently rehydrates necrotic tissue). He was still on the antibiotics and laser treatment, so I hoped that this was just a blip.

During the evening Alba pushed up on his right foreleg to full extension, and tried to push up from the back legs, too. With support from the front he strained to walk forward: he was so determined, and barely two weeks after the accident.

Dinner that night was a hearty

two-and-a-half pounds of chicken, and a litre of water.

13 July (Wednesday)

In a change from the previous day, this morning Alba was a little grumpy, reacting with a growl and snap when his left forefoot was moved. Maybe he was regaining sensation in it, or just in need of more pain relief. After the usual morning routine of cleaning, feeding, meds, TTouch and physio, I opened up an extra run for him so he could move around more, now that he had become so active.

Once I'd done this, Alba came to investigate the new space, pushing himself into a normal sitting position with only the left forefoot knuckled under, which he allowed me to straighten. He sat for about thirty seconds before sliding down onto his side – all unaided. He seemed really happy about the extra room, which was well covered with wood shavings, although his unique way of moving meant he had soon kicked aside the protective layer. Pressure sores were rapidly being replaced by movement sores as his knuckling was causing abrasions on his lower forelegs and front and hind feet, and quickly wearing down his nails. And a hot summer meant I was constantly on guard against fly strike: the kennel, yards, and Alba had to be kept as clean as possible to prevent this.

That evening Alba went one better than he had previously, when he didn't just sit up unaided but very nearly stood up on his own! Using a physio sling to support the front half of his body, Alba stood like this for about a minute, then slumped back down into a sitting position, where he remained for a brief period. When moving, he was definitely using his left foreleg – the worst affected limb – but mostly from the shoulder, in a kind of rolling-it-out-and-flicking-it-forward action. The left forefoot was a concern, still: we knew he had feeling in it, but would he regain movement? As we were only two weeks into the eight-week optimum recovery period, we had reason to be hopeful ...

14 July (Thursday)

The heat of the day meant that Alba was not as active, though he did get himself to the fence that evening to interact with Latea. He tried to lift himself, hindquarters supporting him, and swing the left foreleg forward. Both sisters came over to the fence later, and he was really excited to see them.

15 July (Friday)

Alba had his usual TTouch and physio session first thing, and after the former was able to carry out a full range of movement with both right legs, although the carpal joint was stiff in extension. Work on the left-hand side was limited as he was lying on it. The pressure sore on his left elbow was healing well with new skin growth, but the carpal joint sores looked slightly infected, so it was possible he might require additional antibiotics. Luckily, the vet was coming to see him later, and I lasered them as best I could in the meantime.

Rosie, the vet, arrived at 2pm to check Alba and prescribe more meds. Alba was inside, and more than a little grumpy, so when we went in to look at him, we asked Rosie to remain at a distance. Alba stared and growled at her, so we suggested she move back, but she was having none of it. There was about eight feet, I guess, between her and Alba, and suddenly Alba rushed at her, causing Rosie to beat a hasty retreat: Alba one; Rosie nil!

In a weird way this behaviour really pleased us, as we felt it meant that Alba felt well and strong enough to react to something he didn't like. Even disabled Alba, was not to be underestimated: his majestic power was returning. The vet's heart rate, on the other hand, took a few minutes to return to normal. I again thanked my lucky stars that I was in Alba's good books.

The decision was taken to slightly reduce the painkiller dosage and reassess the situation the following Monday. This might mean that Alba felt sorer, which, in turn, might limit his movement a little. In his condition, there was a fine line between doing too much and doing too little.

That evening, Lunca and Latea were let on to the yard to spend time with Alba again, but only Latea was really interested in staying. Alba excitedly pushed himself around the yard, and, with help, got into a sitting position. When Latea finally went back out into the enclosure, it was obvious that Alba wanted to follow her. Our boy was being so good with staying on the yard, but how much longer would he tolerate it, and at what price to his general health? His nails would soon be worn down to the quick, and the sores on his feet and legs would not improve if he continued to move the way he was doing.

Every day John, Nikki, and I asked ourselves whether it was fair to Alba to carry on, but every day he demonstrated his zest for life, his guts, and ability to fight this terrible debilitating injury. If he was up for the fight, then so were we.

16 July (Saturday): three weeks after the accident, and 17 July (Sunday)

The weekend was not easy for Alba, given his reduced pain relief. He seemed a little listless but that could have been due to the heat. He was starting to become resistant to taking his meds through the syringe, so we decided we'd have to risk crushing the tablets and wrapping the powder in chicken, in the hope that he'd eat it when offered.

He continued to respond to us, coming out of the kennel when called, and we persevered with getting him on his feet. He definitely used the left leg more, and was not dragging it behind as much. Over the weekend he managed to stand unaided briefly – a great achievement – and was observed pushing up into a sit using both forelegs, the left remaining in the correct position for a few seconds before giving out. And, once, he even came out of the kennel using his left elbow to propel himself. It was all very encouraging.

18 July (Monday)

Alba was very affectionate this morning, demanding cuddles, asking for attention, and following me around as I cleaned the yard around him. He took his meds well in the chicken, and allowed me to do a little TTouch and physio, though it was obvious he had other plans this morning, and was being a little mischievous. He did extend his left foreleg against pressure from my hand when I placed it under his foot, remaining in this position until his leg began to tremble. I didn't mind not getting much done with him that morning: it was great to see him in high spirits, and sometimes a break in treatment is necessary – especially with TTouch – to allow the nervous system to process what's been done thus far.

Later, Alba had worked off enough of his high spirits to allow me to laser treat his sores. Unfortunately, those on his lower legs were getting worse, but he simply would not tolerate bandaging, and would have had the dressings off in seconds. All we could do was make the wood shavings as thick as possible, ensuring we repositioned any he had kicked aside.

His earlier activity made Alba a little tired and grumpy again that evening. So far in his rehab he hadn't bitten anyone, which was a miracle, considering the pain he was in, and given his feisty nature. At times like this, the best course of action was to leave him alone to rest and process the day's events – both setbacks and triumphs.

19 July (Tuesday)

Alba joined in the early morning howl session for the first time today, but appeared unsettled after and asked for reassurance. It was almost like he was frightened he'd announced his weakness, unsure whether or not this had put him in danger. As he still generally preferred to be as quiet as possible, the howling was a massive step forward, and fantastic to hear. We'd all missed that lyrical sound as it blended seamlessly with the voices of Lunca and Latea.

His standing exercises were continuing well. Sometimes he seemed to support himself with the hindquarters only, although, little-by-little, was bearing more weight on both forelegs for short periods of time. We continued to use the sling under his chest to get him up, though he was becoming reluctant to accept this.

That evening an alarm call went up from his sisters. They barked, and Alba – who was on the yard with handlers at the time – hurried to the security of his kennel by lifting himself, spinning round, and half-running, half-crawling out of sight. Everyone stared, open-mouthed in amazement, at where Alba had been just seconds before. Fear can be a huge motivator, and Alba had aptly displayed that vital survival instinct by his action. Wolves in the wild do this: den-up to heal if they can; die there if they can't. Pack members stay close by, and will even bring food for the ill or injured.

Later that evening Alba's food was taken to him in his usual metal bowl. He was obviously hungry as he was eager to see what was on the menu, pushing up into a sitting position with both forelegs for a better view, and then up into a full stand on three legs. He sat back down to eat, but when he was lifted up again, took a few steps forward and raked the straw with his right foreleg. Food is also a big motivator for most canines, and Alba was no exception.

20 July (Wednesday)

At 6.30 that morning I found Alba on the yard in a sphinx-like position, but with both forelegs pointing back and out to the side from the carpal joint. He was happy and relaxed, and took water when I offered it to him. His sitting and lying positions were more upright, and he was using his left legs more to get around. He seemed to be using all of the left foreleg, except below the carpal joint, which he knuckled. When asked to, he was able to push against my hand, however.

Whilst I was cleaning, he again proved he was able to get up and move away to hide if spooked, as he ran into the kennel on three legs when his sisters barked at something.

He took his meds concealed in chicken carcasses, and then finished the rest of his breakfast. He was a little too lively to tolerate his pressure sores being treated, moving around more, lifting forelegs as well as hindlegs off the ground. I was able to do more physio and TTouch later, though, and Alba was really relaxed and receptive to the work; truly paying attention to the exercises. Full extension was achieved in the left foreleg, and I was also able to give him a photonic torch and laser session.

Later, after a nap, he was full of energy once more, and I placed him in both sitting and standing positions, although his left forefoot was still knuckling. He was able to take himself to the water bowl: great to see as this encouraged independence and helped strengthen his muscles.

The week continued in the usual way, though we were having to do less to help Alba, and actively encouraged him to try for himself simple things such as moving around, going to the water bowl to drink, and the fence to see his sisters. We continued with his exercises, and he was becoming more and more stable on his legs when standing. We anxiously watched the left forefoot for signs of healing and movement. It was slowly improving, but would it be enough?

23 July (Saturday): four weeks after the accident

One month on, and Alba was about to have a momentous day. Christelle, the physiotherapist, visited again, but, due to Alba's increased mobility, was unable to handle him, so instead demonstrated the exercises she wanted him to do, and coached us from a safe distance.

After getting Alba on his feet I moved each leg in a normal walking sequence a couple of times, to remind the nervous system of the correct order of motion, and help with proprioception (the sense of the relative position of neighbouring parts of the body, and strength of effort being employed in movement). More and more he was able to stand unaided, though slipping on the concrete whilst moving around, meaning his nails had reached a critical point, they were so worn.

Christelle decided that it was time for Alba to leave his temporary hospital area and get back out on to grass. At that time, only two out of the three enclosures at the Trust had what we call holding pens: mini enclosures within the main one, in which to contain the wolves when we needed to cut the grass or carry out maintenance such as hole-filling, tree-planting, etc. The original show enclosure, which the European pack was in then, didn't have a holding pen, so we shuffled around the various packs, placing Lunca and Latea in the main part of the middle enclosure, with Alba in the holding pen. We knew Christelle's suggestion was the right one because, whilst discussing the options, Alba got up unaided and walked!

Although only a short distance, we did have to get Alba down a few steps, but the transfer was accomplished safely and smoothly, and the minute he got onto the grass he stood and cocked his leg to urinate. Of course, we all cheered.

From that point on, Alba never looked back. Our boy – the one the vet advised us to put down just four weeks before, and who all the experts said would take eight weeks to stand, if he ever did – was up in exactly a month. For the rest of the day Alba was unstoppable, and just as when he first began to move in the hospital area, using the roll, slide, roll technique to free his trapped leg, he now came up with a similar manoeuvre, rolling over to get his legs under him, then pushing up with the hindlegs, followed by the fores. He even 'ran' three or four paces, albeit sideways, several times that evening, and showed he could get around the enclosure. To be on the safe side we placed multiple bowls of water around, and cordoned off the concrete area between the enclosure and the hard standing in case he tripped and hurt himself.

The dilemma now was, would he overdo it? I lay awake that night wondering what I would find in the morning. Would he be so sore I wouldn't be able to do his physio? Would he let us near him? Only dawn would tell ...

24 July (Sunday)

The next morning I was delighted to see that Alba was still mobile, and very independent, clearly demonstrating to me that his predicament was *his* problem now, and not ours to help with. He didn't have much movement control, still, and ran sideways instead of walking, stopping only when he ran into something or fell over. And the long grass seemed to be tripping him, so we'd obviously need to do something about that soon. Usually, we let most of the grass in the enclosures grow during the summer, because the wolves loved to hide out in it, making cosy beds and hunting small mammals.

The first few days with Alba in the enclosure were a little tough for me physically, as my email to friends who had been helping with advice and free treatments for him explains –

Hi all,
Just a quick update and thank you to all those who helped and were concerned about Alba, our beautiful wolf with the spinal injury. After his physio session on Saturday, when, growling, he chased out the physio, the decision was made to get him back out on to grass, as the area he

was in didn't provide enough grip (and he couldn't catch the physio!) He had begun to push himself into a standing position, and even attempted to take a few steps.

We moved him into the holding area of one of the enclosures, which is big enough for him to exercise in, but not so big that we could lose him in the long grass. Well, within minutes he was standing, cocking his leg and weeing, and within a day he was running, albeit sideways and with little control. By the second day, after I'd worked on him, he was able to slow his running a little, stop, re-balance himself or let me re-balance him, and then run a bit more.

Wolves learn from a single experience, and this seems to be true for their nervous system as well. I'm going to attempt to bodywrap him in the next few days, and use a form of balance rein to see if he can learn to walk instead of running everywhere, which causes him to lose balance at the moment. And when I say 'run,' I mean run fast: I can't keep up with him. So I'm getting fit this summer chasing a wolf!

Toni x

25 July (Monday)

After a busy weekend, first thing on a rainy Monday we found Alba in one of the day kennels: small wooden structures we fill with straw, which the wolves use in the day sometimes if the weather is bad. Alba hates the rain, which is why he was tucked up in the warm and dry.

He did come out to see us. though, and, because he wasn't interested in food, we had to resort to giving his meds via the syringe, which he wasn't best pleased about. We noticed he hadn't eaten the previous night's food either, which isn't unusual for a wolf. We'd been offering food to him twice a day to aid his recovery, and it was possible he was just reverting to a more normal feeding pattern. During the first month of his rehab he had preferred to eat little and often, which we had encouraged. Was he feeling so much better that he wanted just one meal a day again?

He was obviously cold and sore, though, so I gave him a 45-minute TTouch and physio combined session, after which he got up and ran around for a while. When he halted I placed his legs under him in the correct position (he tended to leave the left fore- and hindlegs out to the side, and let his right-hand side support his weight). His running action was sideways, still – going off to the right – and the left forefoot was sometimes knuckling or getting caught in the long grass. He also seemed to favour his right side more when lying down.

He was much calmer after his usual morning routine. I checked and treated his bedsores (which appeared to be healing from the outside edge in, although sores were developing in different places as his mobility changed: left point of hip, left hind foot, left carpal joint), then left him to sleep for a few hours.

From 12.30 to 1pm Alba ran around energetically. He was still running sideways, so I'd place a hand on his right hip as a guide, and run with him, enabling him to stay straight for a couple of steps before stumbling and falling. It was really hard to keep up with him he was running so fast. When he stopped, I again repositioned his legs into a more normal and balanced stance and then did the physio resistance exercises, which involved gently rocking him side-to-side, asking him to resist the pressure and remain in position, helping muscle development and balance. At rest his left hindleg was generally placed out and to the side of, instead of under, the hip joint.

After repositioning him like this several times, and supporting him when running, Alba seemed to move a little straighter and also whilst standing. Clearly, he was determined to regain full independence, and allowing him to run prevented a build-up of frustration.

The long grass in the holding pen impeded his movement as he was tripping over it, and his left foreleg was becoming caught in it as he dragged the foot. We couldn't take in electrical grass-cutting equipment as this would have upset Alba too much, so the grass would have to be cut by hand with rip hooks – a long and laborious task.

We were concerned that Alba was choosing not to eat red meat, so his feed later was prepared by stuffing mince into chicken carcasses and leaving some trout for the evening shift to feed. The ploy worked; he ate the lot. Mince was the way to go for now, it seemed: perhaps his jaw was still sore?

26 July (Tuesday)

The previous day's activity had tired Aba. We'd spent a lot of time hand-cutting areas of long grass, and, as he definitely moved over the cut areas much better, we really needed to do more. We'd gathered the cut grass into heaps, which Alba appreciated, using them as a warm, soft bed during the day.

He was still willing for me to work on him with TTouch and physio, and even seemed to be helping now, as if he knew what to do. This day he contributed by rolling onto his back and doing full stretches of both hindlegs. He was so funny I couldn't help chuckling to myself. Alba had always been a bit of a clown, and laughing at him always made him act up. His pre-accident character was slowly but surely re-emerging.

We seemed to be reaching a stage where Alba appeared to know what our objectives were, and what we wished him to do. A good example of this was when he willingly chose to eat the beef mince on the menu that day.

27 July (Wednesday)

After his morning meds, Alba was very affectionate, seeking contact. I began a physio session with him but he really wasn't co-operating; just seemed to want to be made a fuss of. Setting off for a walk around the enclosure Alba followed me, his movement much more considered and controlled. He moved more slowly, and had worked out a way to flick his left foot so that, eight out of ten times, it landed squarely on the pad, and didn't knuckle. When he stopped, right fore and hind limbs were under him in a balanced position, although both left legs – although sort of in the right position

– were held out from his side at about 45 degrees.

Alba was very determined to move this morning, preferring to do everything himself. His thoughts were very easy to read because, if I tried to reposition a leg that he wanted to do himself, he gave me a long, hard stare. Eventually, he gave me such a dirty look that I held up my hands and exclaimed "Fine, mate, do it yourself!" He stared at me for a few more seconds, then wandered off, giving the distinct impression that I'd been dismissed.

He did allow me to put him through his shoulder and hip rocks, and would sometimes lean against me for support. But when it came to support when walking and repositioning of his legs, well, Alba was done with that malarky! However, as he was beginning to walk with a curve in his back (his right hip always forward more, with a curve to the right in his spine), to help straighten him and give his nervous system a sense of the correct way to stand, I placed my arms along his sides whilst standing behind him, to help his body balance and align in the correct posture. He tolerated this for one day only, however, before I was summarily dismissed.

28 July (Thursday)

Today, a muscle stimulation machine was tried on Alba, in an effort to counteract his muscle mass loss. It wasn't really successful, however, as he kept moving around, giving us curious looks. The few times we were able to correctly position the equipment, he'd stare, fascinated, as he felt the paddles work.

Alba's rate of recovery was such, now, that we just couldn't keep pace: every time Christelle suggested a treatment, we'd realise he'd already gone beyond that stage, and didn't need it. We knew that wolves have a stupendous ability to heal – research done in the wild tells us that – but to see it happen in real life was something else.

The vet visited again that day, and Alba was taken off his painkillers and antibiotics. As he was getting stronger, it was time, we felt, to see how he did without them.

29 July (Friday)

Alba was in good spirits first thing, suffering no apparent ill-effects from being off the painkillers. Checking his sores I noted that only two (one on his hindfoot and one on the left shoulder) had not scabbed over. Lasering his left shoulder, I could see more skin growth around the outer edge, although the middle looked pussy, and I cleaned it. I was hoping the sun would help with the healing process, too.

Next, I tried some physio, but Alba was simply too active for me to do much.

Alba was now taking a few steps before breaking into the sideways lope; also sometimes trotting a couple of paces. At one point I saw him slowly and deliberately take a step forward on his left foreleg and, with control, place the foot and bear weight on it. He didn't flick it, and I saw him stretch the foot. He appeared to be trying to use that leg more, which made it seem as though his movement was worse, when, in reality, he was using his body in a different way, and improving all the time.

After a while he lay down on his right side, and I was able to properly work the left side for the first time in a while. Alba was relaxed throughout the treatment, and asleep when I left him.

Weeks five to eight, post-accident

Over the next three weeks Alba improved greatly, off all medication and increasingly able to control his movement, although an apparent flick of the left forefoot, and twist to the right of the hindquarters when walking gave him an odd-looking gait. He became more and more resistant to my doing physio on him, but could often be seen extending his legs as if doing his own physio exercises.

By week eight Lunca and Latea were back in with him permanently, which really limited our access to him, of course. We had begun this process by leaving them together in the holding pen in the day and separating them at night; then leaving them together full-time in the holding pen, eventually letting them out together in the main enclosure. Modifications were made to the enclosure so that Alba couldn't fall off platforms, and undo all of the good work so far, but, to be honest, he couldn't jump up on to them, anyway.

He was extremely happy to be reunited with his sisters, with whom he had not lost contact throughout his recovery period. Although the girls took advantage of Alba in play situations, generally, they were gentle and supportive towards him. As he was still unable to jump up on to the bed area of the kennel block, bedding was placed on the floor for him to use.

Over the next few months Alba's carriage improved a little, though it seemed likely he would always walk with an odd, twisting gait, and knuckling of his feet, both front and back. He also appeared diminished when he stood, almost as if he couldn't pull himself up to his full height

As he improved physically and emotionally, Alba began to regain some of his original attitude towards handlers, and many who thought they had built a relationship with him when he was recovering were quickly relegated to again watching him from a distance. Dominance and playfulness returned to their usual levels. He could run at 25 miles an hour – we know this because we clocked him by driving a car past his enclosure one day while we were filming – and loved playing with his sisters, still able to control them if needed, especially if Latea was bullying Lunca. He was still a little unstable; prone to tripping up or being knocked over or aside by Lunca and Latea, who took outrageous advantage of this, and were often seen sitting on him (which he didn't seem to mind, taking it all in good part).

For ages he continued to show a marked preference for chicken carcass rather than red meat, and, as mentioned earlier, we had to wean him back on to the latter by stuffing the carcasses, first with minced beef and then small chunks of beef, until he was finally happy to again eat normal-sized chunks of beef or venison. This aversion could have been due to a reluctance to chew because it hurt to do so. We also took to supplementing his grub by hand-feeding a portion, as the girls would knock him aside as they went into the kennel, and he'd end up with very little or no food by the time he was able to right himself.

EVERYDAY LIFE

By September of 2005 in the colder, damper weather Alba appeared a little stiffer. We were aware that arthritis could be a future consideration, so the decision was made to begin preventative treatment that would support his body, and help with the wear and tear from the over-compensation the original injury caused. Our homeopathic vet, Nick Thompson, prescribed a glucosamine supplement as well as flax oil and multivitamins. By October that year, after six weeks or so on the new regime, Alba appeared to be walking a little better, and standing a little straighter and taller.

Breeding season the following spring was exhausting for Alba, and even though we had suppressed the girls' hormone levels with injections, he still found it difficult keeping the peace that year, not having the stamina to keep up with them as they raced around the enclosure, with Lunca trying to keep out of Latea's way.

Experts are of the opinion that, with a spinal injury, there's a window of about eight to eighteen months in which improvement and recovery are most likely. Over the following months Alba did continue to improve, and even over the next two years still showed small signs of recovery. A number of alternative practitioners offered him healing, which he seemed to enjoy, and from which he did derive some benefit.

The next significant event for Alba occurred a year later when, on 30 Sept 2006, for the first time since his accident, he jumped up onto the greeting platform which stood about three feet high. Until then we had not seen him attempt this, continuing to provide him with a floor-level bed in the kennel block. As his descent from the platform was rather uncontrolled, extra wings – in the form of lower platforms – were added to the sides to help him take the jump up and down in stages.

HYDRO HELPS

That winter, despite Alba's variety of supplements, he was noticeably stiffer. Originally, when it was discussed as an option just after he sustained the injury, Alba's physiotherapist thought that hydrotherapy would be too stressful for him, and the journey to the Greyfriars Spinal Referral and Hydro Centre too traumatic.

However, in January 2007, it was decided that hydrotherapy would be a good option as ongoing therapy for Alba, and on February 15, Alba went for his first session at Greyfriars Rehabilitation Centre. Angela Griffiths, the Centre's owner, had kindly offered free use of the pool on a day when the Centre was closed, so John and I loaded Alba and travelled to Guildford.

Greyfriars had arranged for Carolyn Menteith and Bob Atkins, a writer and photographer for *Your Dog* magazine to meet us to make a permanent record of our success (or failure). We had decided to have two leads attached to Alba's collar to enable us to keep him between us for safety reasons.

Alba was curious, but cautious of entering the building, and hesitated on the ramp leading into the pool. We waited with bated breath. Asking him to go forward we got no response ... but no backward movement, either. Eventually, after a quick conference we tried quiet insistence ... which worked, and, suddenly, Alba was swimming, with us walking each side of him in the shallow pool! We knew he wasn't afraid of water – often going chin-deep into the lake when we walked him with the public – but this was a very different situation, of course: inside, in heated water, with strangers all around.

It became apparent almost immediately that, due to the depth of the water and my lack of stature, I couldn't keep up with the speed that Alba was travelling around the pool. As he seemed calm, we took off the second lead, and then Alba and John were flying. Angela shouted instructions from poolside, and Bob went into a frenzy of photo-taking.

When any animal – let alone a disabled one – swims for the first time, like us, they can tire quickly. In a hydrotherapy pool, a submerged platform on the opposite side from the ramp provides a safe place to rest whilst still receiving the benefits of warm water, which is also weight-bearing, so taking some of the strain off Alba. He didn't swim long that day, just a few minutes, and, as he became fatigued, his left ear began to trail in the water, which he really didn't like. In addition, the injury to his left eye appeared more pronounced. The day, though, was deemed a great success.

A hydrotherapy session became a weekly event for John, Alba, and I. Alba grew stronger and stronger, faster and faster, and could swim for a much longer period. It was noticeable at home, too, that he had much improved movement. At times when he had extended periods off, over Christmas, say, or when we had the foot and mouth outbreak and were restricted with moving the wolves, his mobility rapidly decreased.

I can't claim that Alba ever really loved his pool sessions, but he did take it all with good grace, most of the time. He did, however, love coming out of the pool and standing under the heater. The pool was housed in a poly tunnel, and several patio heaters were dotted around to keep it warm. Once out of the pool, John would hand Alba to me while he went to dry off and change, and Alba would wander around poolside, dripping water as he explored, until one day discovering a patio heater. He stopped and stood under it for a while, then very deliberately turned around to get the warmth on the other side of his body as well. After that day we could have a cup of tea when we were all dried off, Alba laying at our feet under the heater after his swim, just like a dog. It was all very surreal. I know that, to outsiders, John, Alba and I made it look so easy, but we had to warn others to stay clear. Alba did not appreciate unsolicited attention from strangers.

As I was of very little actual use in the pool, I'd hang around the resting platform instead, and when Alba came in for a breather would reach down, under the water, and do some TTouch on him. One very cold January morning after the Christmas break, John and I took Alba into the poly tunnel, which was so thick with condensation, it was like walking into fog. We didn't swim him much that day as he needed building up again after the break, but Alba seemed to really appreciate just sitting in the warm water. He'd go for a lap or two, drift back to me at the platform, have some TTouch, and then, without any persuasion from us, gently drift off for another lap. It was so foggy I could hardly see either Alba or John when they reached the other side. The whole experience took on a dream-like quality, and remains one of the most memorable and magical moments of my life, watching a wild animal, in a hydrotherapy pool, taking control of his own treatment and enjoying the experience with us. We all had the best day ever; I'll never forget it.

Alba swam regularly over the next year or so. The improvement was very noticeable, and he acquired strength from regained muscle mass and mobility throughout his body. At his best Alba was known to do up to 30 or more laps in one session. The hydro also proved a useful indicator to how he was feeling physically, and, from his response in the water, we were able to adjust his medication accordingly.

By the end of 2007 he was quite obviously sorer and stiffer, and a conference with both our regular and homeopathic vet resulted in trialling him on Rimadyl (which reduces joint pain and inflammation in dogs) for ten days to see if there was any improvement. We also tweaked his supplements, and added in homeopathic remedies. Alba improved considerably on the Rimadyl, so remained on a daily low dose that could be increased in the winter or at times when he was especially sore.

During 2008 our visits to the hydrotherapy pool came to an end. Alba had had good and bad days in the pool, but, for a reason we can only speculate about, had gradually become reluctant to enter the building, a disinclination which led to him growling at John and refusing to get out of the trailer.

In October that same year, Alba underwent a general anaesthetic to be castrated, because of aggression levels in the pack. Some years previously Latea had deposed Lunca as the top ranking female, but she wasn't a great boss, and Alba actually still preferred Lunca as a breeding partner. This caused more and more problems, and with Alba unable to fully protect Lunca from Latea's bullying, drastic measures were called for, so we neutered all three on the same day.

Whilst Alba was under we took the opportunity to re-x-ray his back and hips, and bridging (fusing due to arthritis) was found between vertebra T13 and 14, which meant he was able to walk straighter and it was less painful for him to do so. The spondylosis (loosening or breaking down of vertebra) that Alba had developed as a result of his injury was right where his back twisted whilst walking (it's a common problem in older dogs who do agility, or police dogs who jump and twist a lot). The injury site (his neck area) looked good, with no sign of arthritis.

The x-ray also confirmed that we had Alba on the right treatment. When vertebra fuse, it can be uncomfortable until stabilised, so we were ready to increase his pain relief should this happen again. I took the opportunity to get his hips x-rayed at the same time as these were not included in the original investigative images in 2005. There was nothing amiss in this area; in fact, his hip joints would make any dog breeder weep, they were so good.

By January 2009, the cold and damp were affecting Alba greatly, causing stiffness and pain, and harsh winters - snow on the ground and freezing temperatures - obviously did not help in this respect. He was on pain relief permanently now, which a lot of older dogs with arthritis are, although, in the summer, we tended to reduce his dose, until the colder, winter months arrived once again. Eventually, towards the end of that year, we resorted to giving him monthly steroid injections, but could all see he was finding it harder and harder to get around. We took to leaving the kennel open during the day, with a thick layer of straw on the floor for him to lie on, which he seemed to appreciate.

Decision time

By the start of 2010 I became aware of pressure from senior management to make a decision about Alba's future, as his condition was deteriorating. I had assumed the role of head of wolf welfare by this time, after John had left, and although a team looked after the wolves, it fell to me to decide what action to take in situations like this. We'd run out of options, really, but, as was usually the case with Alba, he took the decision out of our hands.

On the night of 22 January, 2010 I ran a Howl Night. The best time to watch wolves is at dusk, or during the hours of darkness, when they are at their most active. On a typical Howl Night I'd give a talk, and then take the attendees outside to see the wolves. Asking everyone to howl their hardest, we would wait, with bated breath, for the wolves to respond. Alba, as usual, gave an impressive display, obligingly vocalising along with his sisters. I remember at one point he was knocked over by Latea and Lunca, but did get back up, even though it was a struggle and took effort. His strange gait was so much more pronounced nowadays: periodically, I had taken short videos of Alba, and viewing these recently, had been shocked at how bad his movement had become compared to older footage.

The next day all of the European wolf handlers planned to give Alba, Lunca and Latea an enrichment walk, because they could no longer participate in the members' walks (Alba couldn't walk that far, and the girls were reluctant to leave him). Those handlers able to work with the pack (all experienced and good mates) would regularly get together for an early morning walk with the three siblings, who loved these enrichment walks – as did we: no public; no trainees to worry about. It was always a special time for us.

That January morning we gathered in anticipation of another lovely walk and bonding time with the Euros. Before starting out, Alba was really unsteady on his feet, and at one point, fell over on top of Latea and couldn't get up. I remember someone asking me whether we should take him, but I thought he was just stiff and would benefit from the walk to work out the kinks.

However, when he collapsed even before we'd reached the field behind the enclosures, we realised there was something seriously wrong with him. His back legs just couldn't support him, and he was unable to regain his feet. Half-coaxing and half-carrying him we returned to the enclosure, at one point trying a supporting towel under his belly which only caused him to growl and snap at us.

Once back in the enclosure Alba tried to move around, but only once managed to regain his feet, and that for a brief period only. I rang the vet; thankfully, he came straight away. Whilst waiting, another senior handler asked what I thought we should do. I hesitated for a heartbeat, then said that I thought Alba had had enough. Another heartbeat passed and then she agreed with me. We looked at each other and, with grief in our hearts and tears in our eyes, went to talk to the others.

As head of welfare the decision, of course, was ultimately mine, but I was grateful to have almost all of Alba's special friends around me to help with this, and so thankful that nobody objected to what I proposed.

GOODBYE, OLD FRIEND

In the enclosure with Alba we said our goodbyes, and comforted each other, and, when he passed, he did so with all of us around him and his sisters close by.

I can hardly type now for tears streaming down my face as I remember those last minutes of Alba's life. He offered no resistance; as with so many times before, he called the shots. We knew it was time to say goodbye to our beautiful boy, and so did he. He was tired: the light had finally gone from his eyes; he did not want to fight on any longer.

Lunca and Latea were nearby, just the other side of the fence, and when it was done and Alba's battered body lay still and lifeless we brought them in to say their goodbyes. They approached cautiously: one of them sniffing Alba's body, and the other going from handler to handler, looking bewildered and seeking reassurance. Finally, the pair moved away, and we carried Alba's body out of the enclosure.

By this time the public had begun arriving, and we'd contained them in the observation room to give us time to grieve in private for a short while before putting on a professional face and getting on with the show. I don't know how the team that was running the walk managed to get through that day ...

I'm grateful that Alba took this difficult decision out of my hands.

The end for him was quick, and with most of his favourite people around him. Now, a few years later, I can look back, with fond memories of the good times. Even though I didn't raise him, Alba – for whatever reason – chose to accept me into his world. Character that he was, the memories I hold dear are of Alba, on a members' walk, cheekily placing his front paws on a handler's shoulders to get a better look at horses passing; Alba and his sisters teaming up to roll a log from the lake bottom to the shore; how he used to go mad with excitement and rush around after paddling, the handlers having to try really hard to keep up with him over difficult terrain; the first time he jumped up and took my whole head in his mouth to greet me, the smell of his meaty breath filling my nostrils.

Most of all, though, I remember his acceptance and tolerance as I helped him recover.

Lunca and Latea crossed over the Rainbow Bridge to join Alba a few years later. Now, in my mind, the three of them run wild and free together through an endless forest, safe from harm; happy to be a pack again.

Raising a pack

In the spring of 2006, the year after Alba's accident, the European pack was officially retired, because Alba was no longer capable of walking very far, and Lunca and Latea were reluctant to leave him. Kodiak and Kenai were not used for public events, which meant that a lot of extra work fell to Duma and Dakota. For a number of years we had been searching for cubs who needed a home to increase our wolf population, but the problem with this is if you don't begin hand-rearing before the cubs' eyes open at around ten days old, they never truly feel comfortable around people. This, in turn, means they're not suitable for ambassadorial work as it's too stressful for them.

Back before birth control was employed with zoo animals, and the introduction of the Dangerous Wild Animals Act, 1976, it was relatively easy to obtain cubs, as there were always litters surplus to requirements, usually resolved by euthanasia: a PR disaster if word got out. I've never been a fan of taking cubs from functioning packs with good mothers but, on occasion, a need to hand-rear presents itself, and one Monday, early in May 2006, John and I found ourselves driving to Dartmoor Wildlife Park to collect three twelve-day-old wolf cubs.

The Park was in crisis at the time, as the local authority had revoked its zoo licence for not complying with health and safety regulations, meaning the Park was closed to the public and operating on a dangerous wild animal licence. The few remaining staff were not being paid, and were using their own money to buy feed for the animals. It was shortly after this that Ben Mee brought the park and reopened it. The story was documented for TV, and later made into the hit film, *We Bought a Zoo*.

The three cubs had had a traumatic start to life. A further two cubs were stillborn or died shortly after birth, and another was abandoned when Lizzie, the mother, changed den sites halfway through labour. The last two spent five days with her, until a rainstorm began to flood the den and Lizzie just abandoned them, too. They were rescued just in time by keepers: cold, wet, and near death. Lizzie was nine years old and this was her first litter: clearly, she just didn't know what to do. Add to this the current instability within the pack (the wolves had begun to fight, and Lizzie became injured), and it was obvious there was no way the cubs would survive unless hand-reared.

The situation at the Park meant the cubs could not be looked after there and, somehow, via the Anglian Wolf Society (a similar organisation to the UKWCT), we were suggested as an alternative carer. AWS had offered us a male cub from a litter bred there that year, retaining the females from the litter for ambassadorial work. It seemed we would have a new pack of four wolves.

After quickly clearing all of the paperwork with DEFRA (the government department for the environment, food and rural affairs), we arrived at the Park. It was an odd encounter – the staff eager for us to be on the road: something to do with the Park's owner wanting to keep one cub as a pet – and we were keen to get started on the long journey home. I was driving my car, with the cubs in a box in the back, cosied up to a wrapped-up hot water bottle. The journey would be tough for them, and they would be late for a feed by the time we got back. Stopping once to check on them, we found that, thankfully, they were fast asleep.

Homecoming

Once back at the centre we quickly transferred the cubs to the artificial den we had prepared. They were very hungry; all three tried to suckle my legs and thumbs as I sat in the warm, den-like area with them. Other staff members rushed to make up bottles of milk. I was exhausted after a long day driving in wet, difficult conditions.

Now I had the opportunity I could see the differences between them. Not yet named, I made up names to remind myself who was who. One had two white front feet so became Two Socks (later named Mosi), like the wolf in the film *Dances with Wolves*; another – the biggest – had a white chest, so I called her Bib (she became Mai), and the last was so small I called her Tiny (this was Mika).

The cubs were surprisingly active, climbing all over me and one another, and Bib/Mai was brave enough to even begin exploring the area. All the while the cubs emitted a small squeaking noise, which they do from birth; as they are unable to hear until they are around twelve to fourteen days old, this can only be for mum's benefit – a "We are near, happy and safe" signal. A very different sound – an urgent squeal – is used if they feel isolated, whilst a yelp signifies they are hurt.

When the bottles were ready I fed Two Socks. All made slurping noises as they drank, and were so hungry we had to take care they didn't take in too much air as they sucked. Cubs paddle with their feet when suckling, an automatic response intended to stimulate milk flow from the mother's teats. However, if they're not correctly positioned when feeding, they can inadvertently scratch with their sharp nails (my arms were constantly sore for about a month). And just as with our own children, we burped them after they'd fed. The problem with bottle-feeding wolf cubs is that they can drink too quickly from the bottle – more quickly than from their mother – so, although they are full, the need to suckle is not satisfied. Wolf Park in the States gives its cubs dummies, I think: I just used my finger!

After burping the cubs, and using a warm, wet tissue to encourage them to urinate and defecate (they weren't old enough to manage this on their own yet), the three of them curled up in a pile next to me and went to sleep. After gazing at them adoringly for a while longer, we crept out, leaving them to recover from the journey and change in circumstances.

I wandered around the site, visiting with the other wolves, all of whom were fascinated by the smell on my clothes, breathing it in deeply. Wolves have a natural instinct to protect their cubs from harm; every one is precious to them, to be cared for, fed, guided, and played with. It is recommended, when introducing adult wolves to new members, that the latter be under three months old, because the majority of bonding between cubs and adults occurs at this time, establishing ties that are lifelong. I've known wolves go mad with excitement when reunited with handlers they haven't seen for years.

New routine

The rest of the day was spent organising rotas. At first it was just senior handlers, and those who had raised cubs before who were included in this, but later others teamed up with us in order to learn the ropes. Consistency is really important when rearing young; you can miss so much with regard to their care if too many people are involved. Pre-weaned animals can decline very quickly, for example: it's always a massive risk and worry. I didn't completely relax until all of the cubs were on solid food, and making their own choices about what they needed to eat. We took copious notes about each cub from day one to ensure everyone was fully informed.

Tuesday, 9 May

The next day, all of the youngsters fed well at 5am, but then things began to go downhill from there. A problem with either the milk or the bottle meant they weren't suckling properly, and struggled to take in the milk. Whilst at the Park, they had been fed by just one person using a very small bottle – was expecting them to suddenly adapt to a standard human baby bottle too much to ask?

The problem was discovered to be the teats on the new bottles, and, thankfully, all was well again by the afternoon. Luckily, this little blip didn't affect the cubs too much, and they gained weight that day (weight gain is obviously a good indicator that all is well, and this should occur daily at this stage in their lives). Of course, different people fed them in slightly different ways, so the cubs also had to get used to that (another reason to limit those involved at first), although with six feeds a day – 4am, 8am, noon, 4pm, 8pm and midnight – it was soon necessary to increase the number of helpers.

Wednesday, 10 May

The male cub, who was to be named Torak, after the main character in Michelle Paver's children's book series *Wolf Brother*, was collected from Anglian Wolf Society (AWS), although, as it was my day off, I didn't meet him until the next day.

Born on April 22 that year, he was nineteen days old to the girls' fourteen. His father was a European wolf, which meant that Alba, Lunca and Latea were related, and could be loosely termed his aunts and uncle, and Mum was a North American. Torak seemed to be trying to be both, as he was twice the size of the girls and very different in colour (grey and brown to their mainly black). He had the long legs of the North American breed but was also stockier, like the European, with a mix of colouring.

Torak was much further on in development, too, clearly able to hear, more alert and active, and responsive when called. He ate a lot more to boot – between 75 and 100 millilitres of milk per feed – and was also able to urinate and defecate without help.

Thursday, 11 May

At 4am the next day I was rostered on to feed the cubs, so met Torak for the first time. He was beautiful and very alert, running straight over to stare at me. He was clearly hungry, and took his bottle very quickly: I fed him first as he was awake and obviously wanting his milk.

The girls were not so quick to wake up, or take their bottles. Mosi took only ten millilitres, and, as I was feeding Mika, she was making noises that Torak came to investigate. He began to howl, then climbed up on one of the straw bales used to construct the walls of the den, and on which I was sitting. Suddenly, I was wrestling two wolf cubs who both wanted to get to the bottle. I could see this one was going to be trouble!

After feeding the cubs I lay down with them. The first time it was just me and them, I was going to enjoy the moment. Peace and quiet: four sleeping cubs and nobody around to disturb my bonding time with them – heaven.

After a while, Mosi began to climb all over me, and started to suckle on my finger – *now* she was hungry and wanted her feed! Eventually, all three girls tucked into my side and fell asleep (one of them snoring), with Torak lying upside down against my leg, twitching. He tried to get on top of the girls' pile but they weren't having any of that as he was so heavy, and rolled him off: they loved lying on top of him, though! Mika developed hiccups as we lay there, and I did my first Tellington TTouch body work on a wolf cub to ease these for her.

Awaking, Torak got up, had a wee and a poo, and brought up a little milk. After I'd cleaned him up he climbed onto my lap and fell asleep again, like a baby, upside down in my arms. He looked so beautiful, his little paws running in his sleep.

All three girls had gained weight in the time they'd been with us, though we'd not yet weighed Torak. He was so big and active, he simply climbed out of the scales when we tried – clearly, another method was needed. In the end, we put him in a cloth shopping bag and hung him from a butcher's scale.

I lay with the cubs that morning til seven. Sometimes they twitched and squeaked so loudly they woke themselves up!

Friday, 12 May

I hadn't seen the babies all day so helped with the midnight feed, when we played and cuddled them for about an hour beforehand so they were all very sleepy come suppertime. I thought they'd begun to recognise me. They settled quickly to their bottles and fed well.

The vet had checked them earlier, and advised we wean the cubs onto rusks and milk first, and then puppy food. (Torak, being a little older than the girls, was already eating some solids.) When weaning it's important that the phosphorus, vitamin D and calcium levels are correctly balanced, as an imbalance can cause all sorts of bone and growth problems. We were advised to use a veterinary-recommended supplement to counteract this.

Saturday, 13 May

At the 8am feed, Torak was tried with

a little beef mince. Mika took less milk than Mosi and Mai, but had had a big feed at four that morning, apparently. All seemed to be urinating and defecating by themselves now, which cut the time it took to feed them.

They seemed very sleepy, this morning, and, after feeding, all four came and curled up with me for a snooze (cubs will always sleep against another body for comfort and warmth, and continued to sleep in a 'puppy pile' for months). Mika had hiccups and burps again, so I winded her, which helped. I also gave her some Tellington TTouch ear work as she seemed very quiet. Stroking ears can have a positive effect on all sorts of areas of the body as there are acupressure points for all in the lobe. It can settle anxiety, boost the immune system, and help sick animals feel better. Mika settled after just a few minutes.

Torak was becoming more and more playful, grabbing at my trousers and shoes. He tried to get the girls to play, too, but he was a little too big for them, and they moved away. I had a bit of rough and tumble with him, which he responded well to. Being slightly older – and having had a much better start in life – meant he was simply too strong and energetic for the girls, which I could see becoming a real problem. Torak needed more exercise and stimulation, which he simply wasn't getting with Mosi, Mai, and Mika.

Sunday, 14 May

Our first crisis today: Mosi did not want to feed, and appeared to be fading a little. She had seemed alert in the morning, although her food intake was down. She would start to suckle but then either fall asleep or struggle with the bottle,and we wondered about calling the vet as she was quiet and a bit floppy. We monitored her closely and she did take some food through the day.

Torak had begun sucking the hair from the girls' elbows and hips, resulting in a slightly moth-eaten appearance. This may have been due to his taking milk from the bottle too quickly, meaning that he still felt a need to suck, or it may simply have been due to boredom on his part. His boisterous behaviour was tiring the girls, and the only solution was for the handlers to play with him more, to give him the exercise he needed. We also decided to contain him in a dog crate at times, to give the girls some peace. Toys also helped occupy him; I brought him a puppy teething ring as all of them were getting their teeth.

Monday, 15 May

I was with the cubs from 4am this morning. Mosi was still not feeding well, though did take twenty millilitres, twice, overnight. She was beginning to look dehydrated, and she didn't want her bottle at four. I spent some time doing TTouch body work on her. After sleeping for a while, she woke at six and took ten millilitres of milk, and again at 8am, but was very slow, appearing dull and lifeless.

We called out the vet, who diagnosed possible teething pain and injected her with antibiotics and anti-inflammatory medication. The treatment worked; by noon she was alert, well, and feeding again: phew!

Torak had his first outing when I took him out of their den and into the education room to meet people. He was very well behaved, if a little in awe, happy to be snuggled into me, under my fleece.

We continued to wean him on rusks, minced beef and puppy food. The girls, being younger, needed a little more time, so we planned to begin weaning them in a day or two.

Tuesday, 16 May

Torak was separated from the girls overnight to give them a break. During the day, we could redirect his energy by playing with him, and I also introduced some more toys for him to bounce off. Later, I took him for another look at the outside world. He seemed to really take it all in, and didn't struggle in my arms. Trips out became a regular occurrence: I'd let him have a little run around in the education room until he began destroying the contents of the shop!

Mosi seemed almost back to normal – feeding really well, and fine apart from a little diarrhoea overnight. Now, it was Mai who wasn't feeding much; probably for the same reason as Mosi: more and more teeth erupting through her gums.

Wednesday, 17 May

All the cubs were fine in the morning. We had started the girls on a little rusk mixed with milk, which they seemed to enjoy, and fed well.

By lunchtime, though, Torak was sick with diarrhoea and vomiting clear mucus. He was also quite floppy. The vet thought it might be a hairball from sucking the girl's elbows, or some other internal digestive problem. Either way, after an injection he was quickly hungry again, and took 160mls of a mix of 50 per cent water and 50 per cent milk. At midnight he took some more milk and solids – puppy meat, rusk and milk. The girls were really eating well but had soft stools: a normal reaction to the change in diet or something a little more sinister?

Thursday, 18 May

Though quiet this morning, Torak still took 90mls of milk first thing, but Mosi was whimpering and had diarrhoea, and apparently straining to urinate. She looked very uncomfortable, her breathing rate twice as fast as the others. She didn't eat, either. By mid-morning Torak had vomited up his breakfast as well.

The vet arrived to check them both. Mosi had a shot for cystitis; Torak another shot of the medication he'd had the day before. The vet thought it could be worms as they were due to be treated (all cubs are born with worms, passed to them via the placenta and mum's milk). We gave them a dose of wormer, plus further doses over the following two days.

Monday, 22 May

I was away for the weekend and, on my return, discovered it had been a difficult time for Torak, who had had to be given lectade to rehydrate him as he wasn't eating. All the cubs now had diarrhoea, due, probably to the wormer they'd had.

Torak had been up and down all weekend, not wanting solids, and this morning he passed live worms, after which he seemed to improve slightly, though was still not keen to take solid food. The vet advised we should see an improvement in the next twenty four hours, and he did become livelier, rolling on his back and playing a bit, though tiring quickly, and appearing a little listless after feeds.

Despite their diarrhoea, all three girls appeared hungry, so, after checking with the vet, we continued with their weaning. They were now lapping up rusk and milk, and even beginning to competition-feed.

Although, of course, we monitored Torak's interactions with the girls, and tried to prevent him sucking their skin, this still occurred. Mai had lost more hair, going bald over most of her body, wih her skin beginning to crack, but we didn't know whether this was due to Torak or the stress of hand-rearing and weaning. The vet advised rubbing olive oil or aloe vera into her skin. As we already had a good quality, organic, edible aloe lotion that we used for Alba, we began applying this to Mai's skin three times a day. The other two girls immediately began licking Mai, so we applied some to them also to distract them.

It seemed that Mosi – now the smallest – was starting to rough-house with the others, and all three girls were more active generally. From a very young age wolf cubs will vie for position within their social group, which can look aggressive, though is normal and without injury.

Tuesday, 23 May

Torak was quickly back to normal, eating solids, playing hard with us, and trying to play with the girls (who were still not keen, being a little too small). All of the girls were eating well, the aloe vera doing its job of restoring hair growth, helped by us monitoring Torak's 'suckling.'

Wednesday, 24 May

Torak remained lively, and even though he passed more worms, these were dead. Mai's hair was growing back, and she really liked the mince that we were now offering with her rusk and milk. Mika was feeding slowly from the bottle, so halfway through I tried her with some soaked rusk. She 'wolfed' down a whole one – wow!

Wolf cubs – like dogs who have erect pinna – start life with the ears flopped over. The girls' ears were becoming erect now, half up, half curled, which looked very comical. They were also more active before and after feeds, and beginning to rough-house with each other.

Torak was chewing on his teething ring more and more, and loved cuddles and games with the handlers. He would have preferred to be with the girls, of course, but was simply too big for them to cope with, so it fell to us to keep him amused.

Thursday, 25 May

At 8am the next day, the team scheduled to feed the cubs were shocked to find Torak apparently close to death: with a low temperature he was staggering, unfocused; mouth agape. He was very limp, and near to unconsciousness. Had he climbed up the straw bale wall of the den and fallen off, they wondered; was he concussed?

Torak was rushed to the vet and placed on a drip, his survival in doubt as the vets were not sure what was wrong with him. Happily, he responded well to the rehydration IV, and was feeding again by the afternoon. A worm count test revealed the level was still very high, despite his being wormed at the weekend, so he was wormed again. We decided to leave him overnight at the vet, just in case he deteriorated.

As pragmatic as ever, Torak coped really well with being hospitalised – even wearing an Elizabethan collar and with IV drip in his leg. After a comfortable night he was feeding well again by morning.

Meanwhile, back at the centre, the girls were going from strength to strength. At just over four weeks old, they were beginning to really play-fight: standing up on each other's shoulders (what we call 'riding up'), growling and grabbing each other around the neck; occasionally pinning each other to the ground. Mai, especially, was really going for it. This was all normal wolf cub behaviour, and it was brilliant to see their personalities develop, and ranking within their group evolve.

They were also competing over food, growling at each other, paddling in the shallow feed bowls as they ate shoulder-to-shoulder; vying for the biggest portion. All were now beginning to prefer solids to the bottle.

Mai's hair was growing back, lighter and very soft, laying flat like adult wolf hair, making her look less like a mixture of hyena/Chinese Crested/Schnauzer. She was the ugly duckling who emerged as a swan.

Friday, 26 May

Torak came home from the veterinary hospital, apparently well, and happily playing with the girls, who – now bigger and stronger – were better able to cope with his boisterous play. Size difference was much less pronounced, now, as the girls were growing at an incredible rate. Mai, in particular, tried to rough-house with him, posturing and growling. The ears of Mosi, Mai and Mika were even more erect.

We had to reinforce the base of the straw bale walls of their den with wood, as Torak kept climbing up and out when we were there as he really wanted to be with us, and whined every time we put him back.

Saturday, 27 May

Mai's ears were officially erect, though looked much too round (I remember thinking, "We really have got a hyena here and not a wolf!"). She seemed set on world domination – even taking on Torak with her play-fighting and dominant behaviour – though, when trying to ride on his shoulders, she kept rolling off because he was so much taller than her. Torak took it all in good part – like a brother with his younger sister – and was such a poppet with her. Every once in a while, though, he would put her in her place, if she overstepped the mark. Mika and Mosi generally let her win and tried to stay out of the way, especially when she was running around with her tail up, growling and trying to dominate by standing over the others.

Mika appeared the least interested in the bottle now, but went mad for the rusk and mince.

All three cubs seemed so much more robust, going from vulnerable and fragile to hardy little monsters, teeth on legs, in the blink of an eye. Constantly trying to chew on me, a telling-off usually got them to back down – all apart from Mosi, who always had to have the last growl.

Deciding to make the den bigger I remember I sat down briefly, and within seconds all three girls had climbed up on me and fallen asleep, though Mika took her time, preferring to play first, nestling around my neck while I made comforting wolf sounds to her. At one point she was upside down in my arms patting my face – very cute – as were Mai and Mosi earlier when they sat facing one another, taking it in turns to pat each other's faces: absolutely adorable.

They were all gaining weight well, looking and acting like real wolves at just over four weeks old (the girls) and five weeks for Torak. As cubs, the girls' muzzles had appeared Pug-like, squashed into their heads, but were now elongating into the adult wolf shape: pointed nose with no discernible forehead.

Sunday, 28 and Monday, 29 May

The cubs were so much more playful before and after feeds now, exploring and playing, and not sleeping as much. They still loved and seemed to need contact with us, though, often laying upside down in our arms and patting our faces. All four were eating well, and often had rusk and milk balls between their toes from standing in and guarding their food. Mika wasn't very interested in the bottle any more, but went mad for solids. Quieter than the others, she still held her own in the rough-housing play sessions, however.

All of the cubs had erect ears now, although very wide at the bottom, and round. Later, these would develop into the small, compact ears of an adult wolf.

Toe Rag (as Torak had now been nicknamed) loved to tug at my trousers, and bite down on my arm really hard, which made me cry out, and he'd stop.

We had noticed that, after feeding, often, they would shiver, though soon stopped when we curled up with them. Was this in anticipation of falling asleep; were they cold if not dried properly after we'd cleaned them?

Mai's hair was regrowing an iron-grey colour; she looked like Spike from *Gremlins* with a Mohican – very strange: maybe she was part-wolf-cross-hyena after all!

Today was my last 4am feed with the cubs as we were cutting meals to five a day at 6am, noon, 4pm, 8pm and midnight. All of them were much more interested in their rusk and mince now, anyway, although Mosi took 115mls from the bottle that morning. Cubs tend to suckle until ten to twelve weeks of age, so we would continue to offer a bottle, from which they took less and less over the following weeks.

Monday, 5 June

(Torak: six-and-a-half weeks old; Mosi, Mai and Mika: five-and-a-half weeks old.) We began taking the cubs out and about for short periods – little walks and play sessions outside the main enclosures but inside the perimeter fence – to mimic natural behaviour in the wild when cubs would leave the den and begin investigating the world. Leads weren't needed as the cubs simply followed where we walked – well, Torak did, he had perfect heelwork, but the girls seemed to go in all directions. They were into everything: climbing in plant pots or chewing on them, eating stones, tunnelling in the cut grass; pouncing on each other and us. We often took them to meet the other wolves through the fence, and all the adults were very interested and kindly: Lunca, especially, was very soft-eyed and gentle with her licks.

The cubs, and particularly Mika, were still squeaking when anxious, but soon this would develop into an alarm bark. Discipline became more necessary as they just wouldn't stop chewing on us. Torak was the best behaved, and, gentle giant that he was, always looked upset when told off. Mosi, on the other hand, was a real feisty character, and would growl when disciplined. All her life she pushed boundaries, and always had a fighting spirit.

The cubs were eating for England, lapping their food really well, so we gave them their first meal without being bottle-fed first, sharing 400mls of milk and 250g of mince, plus egg and rusk.

We planned an outside area for them to be housed in. They needed to fly the nest, and were becoming so active. Mai was definitely in charge of the girls, though Mosi was very bossy. Mika played the role of omega well, and Torak just bumbled along, not having to compete with anyone for the role of top male.

Thursday, 8 to Monday, 12 June

Kenai died on 6 June, which, for me, obviously took the shine off what was happening with the cubs. But it did, however, resolve a space problem, as it meant we then had an entire enclosure for the cubs to be housed in as they grew, as Kodiak could move in with Duma and Dakota.

Though needing time to grieve for Kenai, time and wolf cubs wait for no man, and we hurriedly prepared a small space within one of the holding pens for them to use as a play area. They could then overnight in one of the kennels with bedding on the floor and their dog crate, which they all loved sleeping in.

It was a big week for the cubs, one way and another. Firstly, Mika was diagnosed with cataracts. We'd noticed her bumping into things, becoming left behind, and panicking when in new areas. The vet wasn't sure whether cataracts were forming or dispersing from her eyes, so suggested he monitor Mika over the next few weeks.

The cubs also moved outside to their new accommodation. The play area had wooden day kennels, straw bales, dirt scrapes, and logs for them to play with. Later, we even installed an old Belfast sink for them to paddle in – they loved it! They still slept in the inside kennel, and had the dog crate with warm bedding to curl up in. They all loved this crate, and we'd find all four in it in a pile, fast asleep. It was a bit of a crush, now that they'd grown so much!.

As they could all now lap, we stopped all bottle-feeding, which they were losing interest in anyway, preferring to feed from a bowl. Their dietary preferences were changing also: wanting bone more than milk and mince, and more solid, chewable foods. We still offered milk first, and then mince sometimes with egg and then bone; also some tinned puppy food or chicken wings, etc. All were doing really well on this diet, though – sensibly – weren't keen on the commercial tinned food.

A big problem at the time were crows, who would swoop on the cubs and peck them if they were alone. We resolved this by either someone being there, or, if that wasn't possible, covering the roof of the concrete run between the kennel and the enclosure with a sturdy tarpaulin, to give the cubs a safe and secure area outside.

Sunday, 11 June

Mika's sight seemed to have improved, unless it was just that she was becoming familiar with the new area. She was following more easily, not bumping into things so much, running with her head up, and following our feet and fingers. Although the smallest and obviously the runt of the litter, Mika was always the first to try new things.

Torak was really growing, and could reach up to the top of a straw bale, Around this time he started to be cautious of some people; especially men. He would run from certain volunteers, alarm bark and seem wary, sometimes rallying if the individual talked to him and approached carefully. Other times, though, Torak would run and hide behind me, or try to get through the gate to the rest of the enclosure. We have no idea why he acted in this way, but it may be that someone accidentally scared him, or it could simply be down to his cautious lupine nature: some wolves are naturally more wary than others.

Eating earth – and a lot of it! – became a favourite pastime for the cubs. The earth in this area was clay-based, so, after some investigation, I purchased some edible green clay. Cubs – like a lot of animals – explore everything with their mouths, and I hypothesised that the clay helped settle their stomachs from the new intake of plant, grass and tree matter they had chewed on whilst exploring their new outside area. Another reason might have been that the cubs were instinctively self-medicating against internal parasites: after all, in the wild they don't have access to veterinary wormers! We did re-worm them at this time, too, and Mai passed some worms, though the others seemed fine.

Entering their enclosure meant being met with a really lovely puppy greeting: jumping up, licking faces, and grabbing clothes and hair, before setting off to explore again. Actually laying down with the cubs was the signal for them to check in with you, and, if sleepy, curl up around or on you. When asleep, Torak always sought the reassurance of touching with his nose or paw, and Mai once slept lying between my shoulder blades, her feet each side of my neck. If it was hot or raining, all of them would burrow under me for warmth, or shelter from the sun.

All four cubs had begun to display lots of adult wolf behaviour, such as rolling on smelly food to cover themselves in the scent, food guarding, showing dominance and submission demeanours, and body slamming (wheeling round and crashing hips or shoulders against each other in an effort to knock the 'foe' to the ground). When playing, play bowing, stalking, bouncing and pouncing as if hunting prey was often seen. Whining at them prompted the cubs to come to me, and their own vocalisation was improving with howling and rallying calls, if set off. They listened to the adult wolves a lot, and Torak was fascinated with the Euros, who hung around the fence a lot, looking at the cubs, equally fascinated with him.

Monday, 12 June and following week

Mika had begun to limp badly on her right hindleg the previous day, and this became intermittent on Monday, with her putting some weight on the affected leg. She could easily have strained herself whilst rough-housing and charging around the pen, so we kept on eye on her. Wolves heal at an incredible rate, so an injury you would rush a dog to the vet for – and which might take a long time to heal – can disappear in the blink of an eye with a wolf.

The cubs' eyes were changing colour slowly, going from deep blue to light blue to a bluey orange colour, and in a few more weeks would be the golden amber of adult wolves. The girls' ears were as big as Torak's now, and their feet were huge. All gave the impression of being sturdy, and walked with a steady and assured gait.

Torak continued to be a total delight, climbing into my lap to eat an apple, and allowing me to stroke him whilst he was eating. And he hardly ever chewed on my clothes any more!

More and more wolf-like behaviours were becoming evident, such as ritualised aggression, which is equivalent to us bickering. The loud growls and snarls made it seem as if the wolves really were attacking one another, and biting down, although their teeth rarely made contact with skin, and never drew blood. They also began to show a wolf's advanced problem-solving ability, such as if becoming 'stuck' behind objects, working out how to escape. I saw hunting holds being practiced (holding of hamstrings and necks), and they now used their feet and legs to fend off each other. At times they would 'pack up' against me in play, and I would have to fight them off – just play for now, but was this a sign of trouble to come? When sitting or laying with them, all would approach at some stage to sit in my lap, or, like Torak, sit and lean against me.

They all appeared more instinctively wary of certain things, such as people approaching, and would bark, run and hide until they had identified who it was by their call, whereupon four little heads would pop up from the long grass, and they would hare back to greet whoever it was!

Mika was still bumping into things, though less so, thankfully, which we thought was because she was more familiar with her surroundings. She put up a good fight for her food, and would play with the others. Although the smallest of the four, she was the first to guard a chicken wing!

Mosi's growling when being told off had reduced somewhat, but Mika was growling more! It was as though she was a week behind in her development; her nose was still the shortest, most baby-like. Mika, after all, had missed out on the essential first milk - colostrum - from her mother, which is vital for development and immunity.

They were all growing so much, the girls nearly as tall as Torak, with big ears and paws they needed to grow into. They had strong, straight legs, the special pre-caudual gland, used for intra-species signalling and scent marking, on the base of their tail very evident as white, slightly raised hair: a little like a thumbprint that had been dipped in white paint.

Their diet was modified to include more bone, plus raw chicken wings, tinned puppy food, and chopped up rabbit. We also reduced the mince and milk, and dropped the supplements (cod liver oil, cal-d and gelatine) to one a day on rotation instead of all three daily. With regard to diet, we worked closely with Nick Thompson, a friend of mine, who is a homeopathic vet and nutritional expert. We wanted them to have the best nutritional building blocks.

They were becoming more confident with their vocalisation; not looking to us to ask when the right time was to howl, and answering howls from the other packs.

Monday, 19 June

Shockingly, we lost Mika this day, at just shy of eight weeks old. The previous Friday we had had to rush her to the vet after one of the other girls crashed into her whilst running past. Mika screamed and screamed, and regained her feet holding up a leg: the same one she had been lame on the previous weekend. The vet diagnosed a torn muscle - not too serious an injury - but over the weekend Mika did not improve - she was quiet, not wanting to play and taking herself off to lay down - although she was still eating, moving around, food guarding, etc.

She was x-rayed, and this revealed that Mika's leg was broken and her bone density was very low, making it very likely that more breaks would occur: a little like osteoporosis in women. Even if we could get her over this break, more were very possible, especially during play interaction. There was no cure and no alternative: we had to lose Tiny. She was such an angel, but the odds were stacked against her.

The other girls were blood tested and were fine, thankfully, so it was most likely the lack of colostrum had caused the problem. Either that, or Lizzie somehow knew that Mika wouldn't survive, and changed den sites to give birth to Mosi and Mai. Animals instinctively know so much more than us.

Thursday, 22 June

Today, Torak, Mosi, and Mai had their first whole rabbit, complete with skin and bones. It was gutted to cut down on the worm burden that all rabbits carry in their digestive system, but, apart from a single incision in the rabbit's stomach, essentially, it was given whole. They spent over an hour gnawing at the carcass, and ate about three-quarters of it. They had a little help in the end, as they couldn't quite get into it so we opened it a little more for them. They tore at the meat and crunched at the bone, really enjoying it.

Today, Torak's eyes looked a light amber colour, and the girls resembled panthers – very muscular and powerful. They were growing up fast, and really powering away, both physically and mentally, from how domestic dog offspring would develop. From birth to six weeks of age, the development rate is about the same, but, after this time, wolf cubs mature at a much faster speed.

Friday, 23 June

Their first car journey. When all was said and done, these cubs had a job to do as ambassador wolves, raising awareness and funds for their wild cousins. They would be expected to go into very unusual surroundings and situations during their work, and travel around the country. We needed to ensure they were at least as socialised as a dog, and preferably more so.

I put their dog crate in my car, then John and I lifted the three cubs into the crate. Starting the engine caused a slight panic but they settled when we fed them some mince. We took them on a little journey around the block, during which Mai lay down, whilst Mosi and Torak sat looking out the window. All appeared really relaxed – or so we thought.

Returning to the centre, we could see that Mai was in freeze-mode, her eyes huge. The other two had travelled before in cars (on people's laps) when going to the vet, but Mai hadn't been in a car since I brought all of them home from Dartmoor. (As the healthiest cub, it hadn't been necessary for her to visit the vet.) Obviously, I'd need to give her a few more runs around the block ...

Later, we gave them whole chicken carcasses for the first time, which Mai and Torak loved, though Mosi less so. She took herself off to sleep: it *was* very hot, and it had been a long day, what with the travel training and all.

Tuesday, 27 June

The cubs had their first proper walk on a lead today, which went really well. They were already used to wearing collars, and previously we had attached the leads a few times whilst in the enclosure, but this was the first time we had taken them out around the site.

Torak and Mosi trotted along pretty well, though Mai was more hesitant. We took them to see the Euros, and Latea was very maternal, licking them through the fence, showing her belly: soft-eyed, still and quiet. The cubs loved her and Mai stayed for ages. Torak and Mosi happily trotted off with their handlers, and made it all around the bottom field and back, with Mai catching up after her visit to the Euros. Torak investigated the stream, but Mosi became vocal and anxious, and needed reassurance.

On the way back the adult wolves howled, and Mai looked to us to see if she should howl, now she was out of her usual territory. Wild cubs are not likely to vocalise unless adults do, and will look to them for guidance, not knowing if it is safe to give away their location. When we howled, they rallied round us and joined in with yaps and squeaks. All three cubs were displaying good communication skills.

Saturday, 8 July

(Torak: eleven weeks old; Mosi and Mai: ten weeks old.) The cubs continued to grow quickly, and between them could eat 200g of mince, plus biscuit and egg, each feed. Now on three meals a day, if it was hot they might only eat well in the morning and last thing at night, preferring to miss out their midday meal.

Torak looked the most wolf-like, though continued to be gentle and sensitive; often wary of new people. The girls were still very dark, and looked more robust and stocky. Torak weighed in at 10.7kg, Mosi 7kg, and Mai 7.2kg, so Mosi had really caught up. Both girls were almost as tall as Torak, though not as long in the body.

Mai still wasn't great on the lead; not really interested in going in the same direction as the other two. Latea was still very maternal towards them all when they met, with all of the Euros hanging around the fence a lot.

Today, as I was leaving, the cubs joined in the rally howl, asking me to return. I could hear them screeching and yelping, not proper howling yet, although Torak was the closest to it.

Mid-July

July was memorable for how fast the cubs grew. Every time I looked at them they appeared bigger, stronger, and longer. They were going out for walks regularly now, and were very good on the lead: even Mai was improving in this respect, though still spooked at a lot of things. We took them into the education room, across a wooden bridge, in and out of the trailer, and so on, and only Mai seemed to have problems with any of it. Torak just did it, and Mosi became bold after a few minutes, but Mai took much longer to become socialised with anything. Torak was becoming increasingly shy of some men, and terrified of others. We didn't really know why this was, but, as previously mentioned, there may have been an incident with a male handler, or it could simply be due to Torak's natural wariness.

I heard the snap of Mai's jaws for the first time when she was being picked on by the others from both ends. Their play was increasingly rough and fast, often concluding with one or other leaping into my lap for safety. And they still liked to sneak up behind and grab hair or clothes, occasionally forgetting the rule about not doing so, and having to be reminded with a growl.

Putting them to bed at night was proving a challenge. It was hard enough to catch them in order to get them onto the yard, but getting them in the kennel was even worse: they were like naughty children, avoiding bedtime. We took to putting their food inside the kennel, and opening the door so that they could see it and go straight to it. Once inside, we closed the door behind them. They were never comfortable being locked away, though, especially Torak.

We continued to let them tell us how much food they needed and wanted, and soon they had done away with the midday meal; now on just two a day. They continued to love rabbit, chicken wings and beef chunks, but were slowly going off the mince, egg and puppy food. As it was hot we were also giving them milk to get liquids into them, which they would still take a little of.

Saturday, 29 July

This morning the cubs were very hungry, and between them ate 700g of beef and about eight chicken wings, with Torak eating the most. He had gone through a period of looking very lanky; thin and arched in the back, so we had increased his food and within a week he looked much better. His hair had darkened, too, whilst the girls were acquiring beautiful grey marking on both neck and shoulder, with flecks of white on their faces. Mosi's and Mai's confirmation and movement were simply beautiful: their good bloodlines very evident.

At three months of age I witnessed raised hackles in rough play; also Mai, while being picked on by the other two, displayed a very low tail carriage. Mosi sometimes, now, held hers higher, pushing against Mai's authority. Mai was very responsive and obedient, and would stop chewing on me immediately I growled at her, showing her submissiveness by rolling over. With Mosi, however, it was necessary to be really firm, and even then she still growled, wanting to have the last word.

Torak's lovely nature continued to delight, especially when he wanted to climb onto my lap for a cuddle. He also sought my protection when one or other of the girls got too rough.

All three were howling properly now, without asking us first, and Mai's howl was very deep whilst Mosi's was really high. If I howled with them they would rally round to jump up and hold my face. Wolves muzzle-hold all the time in greeting, when playing, and also when disciplining or appeasing each other. The cubs appeared to like doing this when I howled, especially when they were younger, though I'm not sure why.

Mai was improving on trips in the car, though would wedge herself between my legs. Torak was very inquisitive, and wanted to look out the window, and Mosi paced a bit. Torak and Mosi were often sick, though not Mai.

Of course, the cubs were a big draw for visitors, and we began to inroduce them to work in gentle ways, such as regularly having phorographers in the enclosure with them. Their first walk with the members, and then around the farm with children, went brilliantly, the cubs very confident and happy, and walking well on the lead. Mosi went into the stream, put her nose underwater and blew bubbles, which was very funny. I've never seen a wolf do that before.

Water wonder

One morning in early August, we – me, John and Juliette, another experienced handler – took them out at 7am, whilst it was still cool, and they had a whale of a time. Torak loved the fast-running water in the stream from a recent storm, and splashed up and down in it. They were all chasing and pouncing on butterflies and each other, all wanting to run. Such a memorable time.

Later that same morning we let them out of the holding pen into the main part of the enclosure, and they ran off into the woods to explore and play, finding old bones and sniffing and digging a lot. When it was time to go back we called them and they came running, but, realising we would return them to the smaller enclosure, scampered into the woods again. We had to go into the woods and bring them out on leads!

Wolves are supposed to be anxious about particular noises – Duma and Dakota were very frightened of metallic clatters, for example – but the cubs were mostly bombproof in this respect. Whilst they were growing I put a lot of effort into introducing them to various potentially scary noises and situations, too, and it paid off. Pouring water from a bucket into their shower tray (an addition to the Belfast sink), the cubs loved what I was doing, Mosi even climbing into the bucket with water still in it! When it was empty she pounced on the water I'd poured into the shower tray with her two front feet, and played around with the bucket. All three thought this was great fun, though it backfired on me when they were older, as they would pinch the water buckets and run off with them. We had to chain the buckets to the fence in the end to prevent them doing this.

Sunday, 6 August

The cubs were just gorgeous today: playing with the bucket when I was trying to change the water, jumping into it, picking it up, and listening to me tap the side, their heads turning this way and that to catch the sound from different directions; completely unafraid of the metallic noises.

I even moved the trapdoors on the hard standing up and down with no negative reaction. These sliding metal doors gave the wolves access to the hard standing between kennel and enclosure, allowing us to let them in or out without the need to go onto the yard with them. They're really handy, and essential for safety when working alone, or with less sociable animals such as Kodiak.

I looked in their ears, touched their feet, picked up their legs, and looked in their mouths to get them used to veterinary examinations, with no problems at all.

One time Torak grabbed my foot. There was no pressure, and it was just in play, but when I growled at him he immediately let go, cocked his ears in an appeasing position, licked the end of my nose and walked off: so sweet.

Digging was a new game they'd discovered, especially under the day kennel where their clay pit used to be, and eating whole rabbits was now easy, though Torak ate the most and was filling out more.

Monday, 14 August

Out on a walk, all three cubs voluntarily ate blackberries without prompting, and knew to eat only the ripe ones, too. We regularly gave them fruit and veg such as carrots, apples, blueberries, and greengages but nobody had introduced them to blackberries. Wild wolves often supplement their diet with fruit so this must have been instinctive: wolves are, by nature, opportunistic feeders.

I also observed them eating rabbit droppings and cow pats, all natural and necessary behaviour for health as the droppings provide plant material in a partly-digested form, that the wolves can process more easily than the actual plant, which their mainly acidic gut would have struggled to derive nutrients from.

Saturday, 19 August

We often hid their breakfast in the enclosure to provide them with extra mental stimulation as they sniffed it out. Torak and Mai would rush out, hunting for it, but Mosi usually tagged along behind one of them, and then begged for food once they found it. It had become apparent that, in this respect, Mosi was a follower, not a leader. Today, we hid food on the ground, and also put some in a tree. I was sitting under the tree on a log, and Torak used my shoulders and head as a climbing frame to reach the stash.

When introduced to new things, such as the trailer, Mai hung back like a true alpha, sending in Mosi to investigate (Mosi was bold, and followed Torak wherever he went). Torak was the most inquisitive: jumping up to look in skips, climbing on picnic tables, and hopping without hesitation into the trailer.

Their milk teeth were beginning to change colour, and one of Mosi's canines was grey. This happens when the milk teeth begin to die off as the adult teeth push through. I assumed teething would probably start soon if it hadn't already.

Torak was still the most wary of the trio around people, whilst Mosi would take flying leaps into the unknown: once, when I howled, racing up to me and leaping into my arms. All three still liked to climb into my lap for a cuddle and a fuss. Torak would put his head in my hands, I'd hold my face next to his, and we'd stay like that for ages, communing.

Today, for the first time, I saw Torak marking his food with urine. This was food hidden in a log pile which he couldn't get at, so he marked it in the hope he could claim it later.

Sunday, 20 and Monday, 21 August

(Cubs four months old.) Mai's fear of the trailer persisted, and she refused even to go past it, which made walks very difficult. One time she began to walk past when someone opened a car door, which freaked her out again and she froze.

I tried another tack with her. Instead of reassuring her in the hope that this would allay her fears, I said nothing, and looked the other way. Very quickly you could see her thinking about this, as we watched the others disappear, and, within a short time she got up and walked briskly past the trailer to rejoin the rest. There is a school of thought which says that trying to reassure an animal who is frightened of something only confirms to the animal that they should be frightened, whilst ignoring it completely lets them know that there is nothing to be afraid of.

To build on this initial acceptance – and to confirm in Mai's mind that the trailer was completely harmless – we moved it into the enclosure to help her get used to it. To achieve this, we had to put the cubs on the yard, move the trailer in through the tractor gates, and re-secure the enclosure – it was a mammoth task.

After letting out the cubs afterward, Torak and Mosi went straight to the trailer and began to play by jumping in and out of it. After a few minutes a very relaxed Mai joined them, and, after a few cautious sniffs, was soon doing the same. From then on, all three were jumping in and out, happy as Larry. It didn't take them long, however, to begin chewing on the trailer's cables and rubber fixings, so we had to take it out again! This is typical of wolves, who investigate everything with their mouth and teeth. Still, the exercise served its purpose: Mai no longer feared the trailer, and would happily go inside it.

Next morning, John and I brought the trailer back into the enclosure, and put some food inside. Mai was again very relaxed about it, and ate in the trailer before defecating. We got it out of the enclosure before they had a chance to completely destroy the wiring!

Today was the first time that the cubs met old people with walking sticks and wheelchairs, who came on an organised visit with their club. Torak and Mosi were straight in there, saying 'Hi,' Mai took a little longer to leave the enclosure, but using the same approach of just acting normally, and not trying to reassure her at all worked a treat, and she was soon interacting with the visitors. Walking, she spooked and hesitated a little before the pole barn and cars, but ignoring her while she had a big think about what was going on persuaded her to go past and rejoin the group, stopping to eat blackberries from the bush soon after. The moral of this story is that it's sometimes best to ignore unwanted behaviour and reward that which you do want!

Monday, 11 September

Today, we let the cubs out into the big enclosure by themselves, where they had great fun hunting mice, and jumping in and out of the water trough. They also discovered what fun it was to look through the education room observational window, cocking their heads this way and that when I tapped the wooden frame.

Torak looked very grown up, and could easily be mistaken for an adult. I didn't have to reach down, now, to touch his back, as my fingers rested on it.

September/October

Shedding their milk teeth and eruption of the adult replacements didn't seem to cause the cubs any problems, and the girls were much easier to catch at night, now, thanks to some training. If Torak wouldn't come in, we left him out, which he seemed to enjoy: he never grew used to being restricted to the kennel.

The cubs had become so big and strong that we began treating them like adults, double-leading them in case they pulled us over (having a second lead and handler attached to first for added security when walking).

November

This month the cubs' twice-daily meals were reduced in size a little, with a starve night thrown in, just like the adults. They no longer needed the large intake of nutrients that had aided their growth spurt of the early days. At seven months of age, growth rate now would be slower, and less obvious.

Their behaviour was going through changes, too, becoming much more adolescent, with boundary-pushing conduct, and testing how far they could go with us.

Their coats were really thick, but, apparently, the first year (no matter what the weather) all cubs have thick coats. If the weather the next year is milder, a thinner undercoat will be grown. It seems that, the first year, Mother Nature hedges her bets by providing for the worst case scenario.

December

We regularly rotated the packs in the enclosures to provide new mental stimulation, and give each a spell in the biggest and best enclosure. The cubs went into the bottom pen which, being the newest, was not as mature as the top one, and they totally wrecked it: ripping the sheds to bits, destroying all the wood piles; chewing the saplings. I could see we would have to put together an enrichment programme to give them something else to do!

The 'toys' we provided came in the guise of thick rope, narrow boat fenders, and equine stimulation activities, which they absolutely adored; especially Torak, who became quite good at football, once he discovered he couldn't pick up the boomer balls in his jaws, as they were simply too hard and too big.

January 2007

At night, Torak had begun hanging around with his head just inside the gate, not coming onto the yard. We discovered that if we could catch him and put on a collar and lead, he would quite happily go to bed. It seemed he liked to test us a little ...

Later in the month, we reduced the portion size of their morning meal, giving the main meal at night, which helped with getting them to bed. Breakfast we hid in the enclosure to provide mental stimulation, and watching them locate this was a real highlight of my day.

The next morning the four of them came out of the kennel in a stack, leaping on top of each other in their haste to get through the narrow gap and out into the enclosure. Mai raced off with her tail up and found most of the stash, with Mosi mopping up behind her. Torak was also very good at searching, and really enjoyed digging up the pieces of meat we sometimes buried.

Training this month entailed taking them up the local village hall to prepare for going imto schools, etc. Mai was becoming happier with the trailer, especially if she trusted the person who loaded her. Trailering them the short distance to the hall gave a double exposure.

I thought that Mai would baulk at the hall, but she walked in with me very calmly and quietly, better than Mosi, who, for a while, hung around at the door. Once in the hall, Mai was really good, only spooking a little if things – like chairs – were moved. I didn't think we'd manage it, but, with a little persuasion, and when she saw her siblings there, she trotted up the stairs onto the stage and calmly explored it, only spooking slightly when the curtain was moved.

Loaded in the trailer for the journey home, Torak and Mosi were being a pain, which meant John had to travel with them inside the trailer, as he couldn't get out of the inner cage without them rushing past him.

Torak now weighed in at 37kg; Mai around 28kg, and Mosi the smallest at 27kg.

Wednesday, 24 January 2007

The cubs experienced snow for the first time and loved it, running at full speed, their noses buried, ploughing a furrow; leaping to catch thrown snowballs; jumping on the ice in the water troughs and playing with it. A joy to watch. We even made them snowmen to destroy.

Friday, 26 January 2007

This morning I confused the cubs by hiding their breakfast in the opposite side of the enclosure. When released, they all ran to the left, then very quickly turned and ran at full speed to the other side, with Mai – tail up – in the lead. After breakfast, we went for a walk around the enclosure, and Mai and Mosi were playing with me, making play faces and jumping up. There was no biting, though, and they were easy to control by my placing my hand over their muzzles, or pushing one onto another to divert them.

Later, as I was leaving, I said goodbye to them through the fence. I had a bucket in one hand, and, obviously thinking this held food, Mai had a right pop at Mosi, with Torak joining in, too, to remind her that it wasn't her place to go first. Mai really is a perfect leader, sending in the omega to check out a new situation, but keeping her in her place when need be.

There was still no sign of the girls having a normal breeding season hormone rise: ordinarily, from November on, adults wolves will interact more aggressively with each other, and become more difficult to handle. Torak's testicles still hadn't dropped, either. Wolves are said to sexually mature at twenty-two months of age, but I did wonder if the lack of adults in their group would cause the cubs to mature early. It didn't, however, and it wasn't until the next year that the girls reached sexual maturity and had a season. Torak's testicles didn't drop until February that year, though he didn't seem to mind me checking every now and then ...

Monday, 5 February

I had a lovely time with the cubs this morning, all jumping around and on me while I sat on the greeting platform, although it was getting a little hard to handle them on my own now. Torak came out of the kennel and, instead of running off to find his breakfast, came to me and held my arm in greeting, then led me out into the enclosure wearing a happy 'play' face (equivalent to a human grin).

Later on the platform I had Mosi on my shoulders and Mai and Torak each side, all muzzle-holding me and playing. Really enjoyable though not good for your hairstyle!

Tuesday, 6 and Wednesday, 7 February
Torak stayed out both nights in temperatures that dropped to -7 degrees, so he had frost on his coat in the mornings. Wolves lose very little body heat, and can often have frost or unmelted snow along their backs as a result.

The first morning he was so happy to see the girls, and they charged around playing and skidding on the frosty ground. A real joy to watch. The next morning I let Torak on to the yard to meet the girls coming out, and he stole a chewed-up rabbit skin from them for his breakfast, which kept him quiet while I went off for a walk around the enclosure with Mai and Mosi, following me when I called them. We had fun breaking the ice on the water trough, and licking the frost-covered grass. Mai was very affectionate and jumped up for a cuddle – a lovely way to start the day.

Growing up

The Brat Pack, as I came to call them, continued to mature into their roles of ambassador wolves at the UKWCT, though still retaining their individual characters and idiosyncrasies. Torak was still bold as brass around objects and buildings, and was often called upon to do photo shoots standing on diggers, or being filmed on green screens (where the background is added later) for TV. He remained wary of men, however, and became increasingly hard to catch if he knew the public or his most-feared male handlers were on-site. Loyal to those he loved, he no longer works with the public, though enjoys going out on enrichment walks with his favourite people.

Mosi remained a pushy little madam, always testing the boundaries and scaring people half to death by 'talking' to them (a low grumble or growl) as they stroked her. I used to prove to folks she didn't mean it by walking up to her and kissing her on the head while she was in full growl. Eventually deposing Mai as top female, she and Torak now live together, separate from Mai.

And Mai, my sweet and gentle, people-loving girl, got a new mate. Motomo was brought in from the wolf centre in Devon to live with her, and they just adore each other. Mai's jet black cub coat turned prematurely grey over the years, and she is almost as white as her mother, Lizzie. Mai became a mother herself to the Beenham pack (born after I left the Trust).

And that's the story of raising a pack. It nearly turned me grey in the beginning, worrying about them, and going through a few too many crises before they were weaned, and losing Mika so young was heartbreaking.

What did I gain from the experience? A wealth of knowledge about wolf behaviour; insight into the difficulties most wolf parents must go through to raise a pack in the wild, and a whole host of joyful memories of my 'babies' who, to me, will always be Bib, Tiny, Two Socks and Toe Rag ...

Over the next three years, once the cubs had grown and were settled into full-time work at the UKWCT, other things took priority. Alba, Dakota and Kodiak needed my attention more, either because they were older or their health had declined, and I helped nurse each of them until their deaths.

Life took on the ebb and flow of the changing seasons, and people and events came and went. Under my care the education programme expanded as much as it could within such a small organisation, and I began to fill restricted, and even bored at times, running the same events, year-in, year-out. My work commitments outside of the Trust were beginning to suffer as I spent more and more of my time with the wolves.

And I felt the need for a new challenge.

Moving on

Consequently, in August 2010 I resigned as education officer at the UK Wolf Conservation Trust. We had created this role when Roger died to enable the week day education programme and events to continue. Having achieved this, I felt I needed to get back to my business, which had been sorely neglected over the last few years. I also wanted to write.

Another reason for leaving was that an amazing opportunity had arisen to work for an agricultural education charity called The John Simonds Trust, an award-winning organisation that works out of an organic farm, just one valley over from the wolf Trust. I was excited about learning new skills, and expanding my knowledge to encompass farm animals. As the education programme at Rushall Farm ran for only half the year, in the growing season, the position also gave me the time to devote to my ongoing Tellington TTouch Training Practice ... and I would finally have time to write. My first book, *The Truth about Wolves and Dogs – dispelling the myths of dog training*, was published by Hubble & Hattie in November 2012, and met with great success.

Wolf contact

I was heartbroken to leave the wolves, of course: wolves are addictive, and having no contact with them would be difficult in the extreme for me. The worldwide wolf community is relatively small, however, and, thankfully, as I had made many friends around the world and in the UK, still talk to wolf biologists, and handlers and keepers of captive wolves. Offers to visit wild wolf projects are open-ended and, in the UK, trips to see friends with socialised wolves are a must.

For example, I regularly visit my good friend Pia Gismondi, who works at Paradise Wildlife Park. Pia and I worked together at the UKWCT before she followed one of Apollo's and Luna's cubs to the Park. Paradise has two packs of wolves: two older, socialised females whom I often get invited into the enclosure to interact with, and a part-socialised pack of three Norwegian wolves, who I hand-feed choice cuts of meat through the enclosure fence: pure heaven for me!

These rare and special moments lift the spirits and inspire

me. The emotions I experience when a fully grown wolf approaches, sniffs, and then delicately jumps up to place her paws on my shoulders for a wolfie version of an intimate greeting are indescribably wonderful, and I live for these moments.

I also remain in contact with Caroline Elliott from the Anglian Wolf Society, Anne Riddell from Wildwood Trust in Kent, and Tony Haighway and Deb Seward from Wolf Watch UK. People who raise, manage and care for wolves are few and far between: members of an exclusive club in which we share our knowledge. If there's a problem, we talk, and I often receive a call or email asking for my opinion on how to handle a particular procedure or situation. I'm more than happy to help, and will continue to do so to make the lives of these captive creatures as fulfilling as possible.

The ethics of captive animals

Most people could be forgiven for assuming that I'm pro-captive animals when, in fact, I'm not. In my perfect world zoos would look very different, and even not exist at all. The recent advances in the study of animal emotions gives even more credibility to the argument that certain species – those who exhibit autonomy, self-awareness and emotion, for example, and who forge strong social bonds, should be afforded the status of personhood (the cultural and legal recognition of the equal and unalienable rights of human beings). This has led to legal arguments in support of non-human animals, and prompted American lawyer Steven Wise to create the Non-Human Rights Project (http://www.nonhumanrightsproject.org/steve-wise/), an organization working through the common law to achieve actual LEGAL rights for members of species other than our own. Wouldn't it be interesting if, in the future, both captive, domestic and wild animals were afforded these rights?

But even without personhood status, there can no longer be any reason to doubt that ALL animals are sentient beings, capable of many of the same senses, feelings and fears as ourselves. Highly intelligent and emotionally aware, non-human animals are daily subjected to painful and traumatic experiments in labs (often without pain relief); made to perform unnatural behaviours, or held captive in tiny spaces, devoid of any and all enrichment, purely for our viewing pleasure. We control every single aspect of their lives: what they eat and when; when they sleep; if and who they mate with; how and if they get to rear their young, and when the young are taken from them. Animals who would naturally live their entire lives in company with their mother are taken from them when young, like the Orca calves who are snatched from the wild, or captive creatures who are moved to other facilities. Some animals are culled, deemed surplus to requirements just because they are male. We do terrible things to animals every day.

And wolves don't fare much better. If captive but not socialised to humans, they remain fearful of us their whole lives, employing stereotypical behaviour such as pacing to enable them to cope with the stress, or simply hiding away (though are often housed in tiny enclosures with nowhere to hide, so that they are permanently visible to the paying public). In the wild, a pack's territory would comprise hundreds of miles, over which its members would freely roam. And cramped accommodation also means that young, captive wolves are unable to disperse, which causes friction in the pack and in-fighting. They cannot be relocated to a zoo as it's not possible to mix adult wolves with other adult wolves, unless it's two individuals in mating season who could start a new pack.

Wolves breed really well in captivity, and, as a result, we have too many. In the old days, before reliable contraception was introduced, cubs would simply 'disappear,' with the lucky ones ending up at wolf parks like the UKWCT to be hand-reared.

If a wolf cub is abandoned by his or her mother, I can justify in my mind the need to hand-rear the animal. After all, socialised wolves (a wolf habituated to human contact) have a much less stressful life in captivity than unsocialised wolves, who will never feel comfortable in the presence of humans. Checking their health regularly, and treating any illness or injury is easy to do; mental stimulation can be provided, and they can even go for walks. But taking cubs from their mother when barely a week old to hand-rear them as socialised wolves and public exhibits is reprehensible.

Too much commercial value is placed on animals, especially those who can draw big, money-spending crowds. Take Tilikum, the Orca at SeaWorld in Amerca, who, during his life as a captive animal, has killed three people. A domestic dog kills one person, or just looks like the 'wrong' breed, and he or she is instantly euthanised. Tilikum lives because his sperm is needed: he has something that man wants. He now resides in near isolation in a tiny pool, floating motionless for hours on end, a living sperm bank. He deserves better – all animals deserve better – and we must fight to achieve this.

My hopes for the future

Research shows that zoos contribute little to educating the public about animals in the wild. People go to zoos to entertain their children, not learn about the plight of animals. Think about it, can you remember one fact you read or heard while visiting an animal park? Did you rush home and donate to a conservation charity, or start a campaign to make the lives of animals, safer, better, and healthier? The answer for the majority is probably 'no.' I don't have the answer to what a zoo should look like in the future. if they are to remain, but I know they cost a lot of money to maintain. Could that money be better used to make a difference elsewhere? Of course it could. Most zoos will support wild projects, but the proportion of profits that gets to the front line is minimal in comparison to what it costs to run them.

My biggest concern is that zoos should have to change their management techniques to include not just the physical but also the emotional needs of the animals in their care. Mental stimulation and environment enrichment are words you will hear a lot from zoo keepers, and, yes, some try to keep their wards entertained to stave off boredom and promote natural behaviours. However, this needs to evolve to include emotional wellbeing. Animals should not be seen as commodities, 'things' that can be moved around or destroyed in the name of genetic diversity, guarding against extinction, or monetary gain. The stress that zoo animals experience in their lifetime we can only estimate, but it will be huge for some.

My heartfelt hope is that public awareness will force change, but whether that will be in my lifetime, I just don't know. In the meantime, I can do my bit by, hopefully, encouraging you to think differently about how we treat animals, and what their lives are really like.

What you can do

Support only the good zoos: responsible places of education and science, whose prime concern is conserving the wild; who house animals in the social groups and habitats they like and need. Some zoo managers are recognising that the way forward is to have specialist centres for just one type of animal, where everyone understands the needs, behaviour and health issues of that species.

A good example of this is Monkey World in Dorset, England, a purpose-built centre with up-to-date enclosures and plenty of space. Accommodations are large and enriched, staffed by people with excellent knowledge of the species. The monkeys come to the sanctuary from all over the world, rescued from appalling conditions, and nursed back to health. Breeding is limited to endangered species. Education about the plight of these animals is high on the list at Monkey World, backed up by an excellent TV series that shows the work behind-the-scenes.

Places such as Monkey World carry out invaluable work, which is generally well-received, if donations are anything to go by.

Of course, directly donating your well-earned cash to a wildlife project is another excellent option, but choose a good one, and research how it operates. It's a sad fact that conserving a species whose habitat is disappearing won't do a great deal, and the same is true of a charity that does not take into account how the local community – usually shockingly poor and in dire straits – can be helped also. Many NGOs (Non-Governmental Organisations) do brilliant work, but ask for facts and figures. For example –

- What percentage of donations get to the front line projects out in the field?
- Who runs the project, are locals involved, and at what level, if so?
- What is being done to conserve the habitat of the animal concerned? Trying to save the Giant Panda when the animal's staple diet – bamboo – is being destroyed will not be successful
- Does the NGO have a standing with local government, and access to worldwide policy-making?

- Is the work it does relevant to the area and animals in it?

It's another sad fact that no wildlife conservation project will succeed unless the human aspect is also considered. For example, wolf conservation is pointless if the animals are then trapped and killed by farmers to prevent them killing their livestock. Far better (and ultimately more beneficial) to first provide farmers with fencing for their livestock, and herd protection dogs so that the wolves cannot get access to the livestock. It might not be fair, and it might not be reasonable, but the world will always expect that our own species takes precedence over others, although we must do what we can to ensure equality, if at all possible.

My fervent wish is that my book will motivate *you* to do something for your favourite animal, be that animal wild or domesticated. Speak out for animal rights, learn as much as you can, fund raise if you have the means to, research if you are in that field, write about it.

Do whatever you can: together we can make a better future for all animals.